Performance and controller design for the synchronization of multi-agent systems

Dissertation zur Erlangung des Grades
eines Doktor-Ingenieurs
der Fakultät für Elektrotechnik und Informationstechnik
an der Ruhr-Universität Bochum

Andrej Mosebach
geboren in Ananjewo, Kirgistan

Bochum, 2017

1. Gutachter: Prof. Dr.-Ing. Jan Lunze
2. Gutachter: Prof. Dr.-Ing. Ulrich Konigorski

Eingereicht am: 31.03.2017
Tag der mündlichen Prüfung: 21.09.2017

Bibliographic information published by the Deutsche Nationalbibliothek

The Deutsche Nationalbibliothek lists this publication in the Deutsche Nationalbibliografie; detailed bibliographic data are available on the Internet at http://dnb.d-nb.de .

ISBN 978-3-8325-4623-6

Logos Verlag Berlin GmbH
Comeniushof, Gubener Str. 47,
10243 Berlin
Tel.: +49 (0)30 42 85 10 90
Fax: +49 (0)30 42 85 10 92
INTERNET: http://www.logos-verlag.de

Acknowledgements

This thesis is the result of my work as a research assistant at the Institute of Automation and Computer Control of Prof. Dr.-Ing. Jan Lunze at the Ruhr-University Bochum.

Firstly, I would like to thank my advisor, Prof. Dr.-Ing. Jan Lunze for his continuous support and motivation during the past five years. This thesis would not have been possible without his immense knowledge that he shared with me during our discussions. I also thank Prof. Dr.-Ing. Ulrich Konigorski for accepting to review this thesis and his comments regarding the manuscript.

A special thank goes to my former colleagues at the Ruhr-University Bochum: Alexander Schwab, Christian Stöcker, Christian Wölfel, Daniel Lehmann, Daniel Vey, Fabian Just, Kai Schenk, Markus Zgorzelski, Melanie Schuh, Michael Schwung, Ozan Demir, Philipp Welz, René Schuh, Sebastian Pröll, Sven Bodenburg, Tobias Noeßelt and Yannick Nke. I have made good friends at this institute and I am grateful for the moments I have shared with you. Besides my colleagues, I would like to thank my student assistants Christoph Kampmeyer, Evgenij Polyakov, Simon Röchner and Svenja Mozian.

I want also to thank Andrea Marschall, Kerstin Funke, Susanne Malow, Rudolph Pura and Udo Wieser for supporting me with the administrative and technical work.

My deepest gratitude goes to my family for their love, understanding and the unconditional support during this challenging time.

Finally, I want to thank my wife and best friend, Viktoria, who has been and will always be my inspiration. Thank you also for giving birth to our beautiful daughter Laura.

Bochum, October 2017 Andrej Mosebach

Contents

Abstract

The analysis and design of control strategies for the synchronization of subsystems that are coupled over communication networks is the topic of this thesis. Typically, synchronization problems deal with the asymptotic behavior of networked multi-agent systems, where it is required that the states of the subsystems follow a common trajectory as the time approaches infinity. In contrast, this thesis focuses on strategies that do not only fulfill the requirement on asymptotic synchronization but also requirements on the transient behavior of networked multi-agent systems. Motivated by a growing number of applications where subsystems exchange their information by means of modern communication systems, the limits on the achievable performance of synchronization are studied for large teams of autonomous subsystems. In particular, control strategies that do not require any centralized coordination of the subsystems are developed.

Two types of performance specifications are introduced with respect to the transient behavior of the overall system. Specifications concerning the transient behavior with respect to some time interval and specifications concerning time instants like the requirement on some settling time. To reflect the overall system behavior, an infinite time horizon linear quadratic objective function is chosen which penalizes the state differences between subsystems that are close to each other. Furthermore, requirements on the settling time of some overall synchronization error and the damping of oscillations are introduced. According to the mentioned performance specifications, three different design approaches are proposed for the synchronization of networked subsystems by static state feedback controllers, dynamic output feedback controllers and for the synchronization of subsystems communicating over point-to-point connections.

The design of static feedback controllers is achieved by introducing an optimization problem for the overall system, which penalizes the structure of the communication network. A networked controller which solves this optimization problem fulfills the requirement on optimal synchronization. It is shown that optimal synchronization is typically computationally expensive to achieve. For this reason a gradient descent algorithm is proposed that ensures optimal synchronization for any given state feedback controller satisfying the weak requirement on asymptotic synchronization.

The key idea for the design of synchronizing dynamic output feedback controllers is based on the creation of a dominant pair of poles in the dynamics of some decomposed synchronization errors. This procedure leads to a transient behavior of the synchronization errors, which is bounded with respect to the algebraic connectivity of the communication network. This bound shows that the algebraic connectivity has a significant influence on the overall behavior. The main idea of the controller design procedure is to translate the requirements on the synchronization time and the damping into specifications of the bound, which are afterwards used for the determination of the controller parameters. Finally, the design approach is extended by the use of homogenization methods for the synchronization of multi-agent systems with non-identical dynamics and directed communication networks.

The synchronization analysis of multi-agent networks where only point-to-point connections are allowed reveals the following two interesting properties. First, asymptotic synchronization of the overall system can only be achieved if time varying couplings among the overall system are introduced which ensure synchronization of the temporary coupled pairs of subsystems. Second, the analysis of the system behavior shows that asymptotic synchronization of unstable subsystems can only be achieved if the synchronization between coupled pairs is faster than they divergence according to their unstable behavior. Based on this properties two control algorithms are derived which ensure not only the requirement on asymptotic synchronization of deterministically or randomly changing pairs of subsystems but also the requirement on a settling time of the synchronization errors.

The applicability and performance of the developed design approaches is demonstrated in simulations and by experimental data obtained from the synchronization of mobile robots.

Kurzfassung (German abstract)

Das Ziel der vorliegenden Arbeit besteht in dem Entwurf und der Analyse von Regelungsverfahren zur Synchronisation von autonomen Teilsystemen, die über ein gemeinsames Kommunikationsnetz verbunden sind. Überwiegend verfolgen Methoden zur Synchronisation von Multiagentensystemen das Ziel, die Teilsysteme asymptotisch auf eine gemeinsame Trajektorie zu führen, ohne dabei das Übergangsverhalten zu berücksichtigen. Diese Arbeit konzentriert sich im Gegensatz zur Literatur auf den Entwurf von Regelungsalgorithmen, die zusätzlich zur Forderung an das asymptotische Verhalten auch Forderungen an das Übergangsverhalten berücksichtigen. Im Mittelpunkt dieser Arbeit steht der Entwurf von Methoden, die keine zentrale Komponente benötigen, um das Synchronisationsproblem zu lösen.

Es werden zwei Arten von Gütekriterien, die sich auf das Verhalten des Gesamtsystems beziehen, vorgestellt. Gütekriterien, die das Verhalten des Gesamtsystems bezüglich eines Zeitintervalls abbilden und Gütekriterien, die sich auf einzelne Zeitpunkte beziehen, beispielsweise die Forderung nach einer Übergangszeit. Zum Beschreiben des Systemverhaltens über den gesamten Zeitbereich wird ein lineares quadratisches Gütefunktional verwendet, welches nur Zustandsdifferenzen zwischen benachbarten Teilsystemen gewichtet. Des Weiteren werden die Dämpfung und die Überschwingzeit des Synchronisationsfehlers, der sich auf das Gesamtsystemverhalten bezieht, als Forderungen verwendet. Für die genannten Gütekriterien werden in dieser Arbeit drei Reglerentwurfsverfahren zur Synchronisation von vernetzten Teilsystemen vorgestellt. Dabei können die Teilsysteme über statische Regler, dynamische Ausgangsrückführungen oder über Kommunikationsnetze, die nur Punkt-zu-Punkt Verbindungen erlauben, vernetzt sein.

Der Entwurf von statischen Reglern beruht auf einem Optimierungsproblem, welches sich auf das Gesamtsystem bezieht und das Teilsystemverhalten bezüglich der Struktur des Kommunikationsnetzes gewichtet. Die als Lösung des Optimierungsproblems resultierende vernetzte Regelung erfüllt dabei die Forderung nach optimaler Synchronisation. Es wird gezeigt, dass vernetzte Regelungen, die diese Forderung nach optimaler Synchronisation erfüllen, in der Regel nur rechenintensiv bestimmt werden können. Um die Komplexität zu reduzieren, wird ein Gradientenverfahren vorgestellt, mit dem eine gegebene asymptotisch synchronisierende vernetzte Regelung so verändert wird, dass sie zusätzlich die Forderungen nach optimaler Synchronisation erfüllt.

Die Idee für den in dieser Arbeit vorgestellten Entwurf von dynamischen Ausgangsrückführungen besteht in der Erstellung eines dominanten Polpaares in den Übertragungsfunktionen entkoppelter Synchronisationsfehler. Dieses Verfahren führt zu einem Übergangsverhalten der Synchronisationsfehler, welches im Bezug zur algebraischen Konnektivität des zugehörigen Kommunikationsnetzes beschränkt ist. Ausgehend von der geforderten Überschwingzeit und der Dämpfung werden Forderungen an eine obere Schranke und die daraus resultierenden Reglerparameter bestimmt. Unter Verwendung von Homogenisierungsansätzen wird gezeigt, dass sich dieses Entwurfsverfahren auch zur Regelung von nicht identischen Teilsystemen und von Kommunikationsnetzen mit gerichteter Topologie eignet.

Die Analyse von Kommunikationsnetzen, in denen Daten nur über Punkt-zu-Punkt Verknüpfungen ausgetauscht werden können hat zwei interessante Zusammenhänge aufgezeigt. Zum einen lässt sich asymptotische Synchronisation des Gesamtsystems nur dann erreichen, wenn unterschiedliche Paare von Teilsystemen immer wiederkehrend für den Synchronisationsvorgang verwendet werden. Zum anderen kann asymptotische Synchronisation instabiler Teilsysteme nur dann erreicht werden, wenn temporär verkoppelte Teilsystempaare schneller synchronisiert werden als sie divergieren können. Basierend auf diesen Zusammenhängen werden Entwurfsalgorithmen für ein deterministisches und ein zufälliges Verkoppeln der Teilsystempaare vorgestellt, mit denen nicht nur die Forderung nach asymptotischer Synchronisation, sondern auch die Forderung nach einer Übergangszeit erfüllt wird.

Die Anwendbarkeit der vorgestellten Entwurfsmethoden wird anhand von Simulationen und Experimenten zur Synchronisation von mobilen Robotern demonstriert.

1 Introduction

1.1 Synchronization of multi-agent systems

Synchronization is a widespread phenomenon appearing in groups of interacting individuals that share a common goal. Vehicles driving in a platoon, mirrors focusing the sunlight in a solar power plant or rotating rolls in a paper converting machine represent only a few technological examples where autonomous subsystems have to reach an agreement to achieve a common control goal. The corresponding overall systems that are composed of two or more physically uncoupled subsystems are referred to as multi-agent systems.

The information exchange between the subsystems, which are also known as agents, is essential for synchronization. Synchronization of unstable agents towards an unstable trajectory can only be achieved if couplings within the overall system are introduced through which each agent is influenced by every other agent directly or indirectly. A complete coupling of the multi-agent system is often desirable and gives the maximum degree of freedom for the design of synchronizing controllers. But even when modern communication technologies are used, there typically exist restrictions in the interconnection of the agents that are caused by limited communication or energy resources. These restrictions lead to various interconnection topologies, each of which influencing the behavior of the agents in a different way. As long as the topologies include a spanning tree within all agents, asymptotic synchronization is possible. However, the goal in this setting is to find methods for the design of synchronizing controllers which do not only ensure the requirement on asymptotic synchronization of the agents but also specified performance requirements for a given interconnection topology.

This thesis presents methods for the design of local controllers $\Sigma_{\mathrm{C}i}$, $(i = 1, 2, \ldots, N)$ for the synchronization of agents with identical dynamics and a given communication network with the set-up shown in Fig. 1.1. The communication network together with the local control algorithms is referred to as the networked controller. In particular, this thesis focuses on methods for the design of synchronizing controllers with respect to the transient behavior.

The agents are described by general continuous- or discrete-time linear state space models with the states $\boldsymbol{x}_i(t)$, the inputs $\boldsymbol{u}_i(t)$ and the outputs $\boldsymbol{y}_i(t)$ for $i = 1, 2, \ldots, N$. In order to

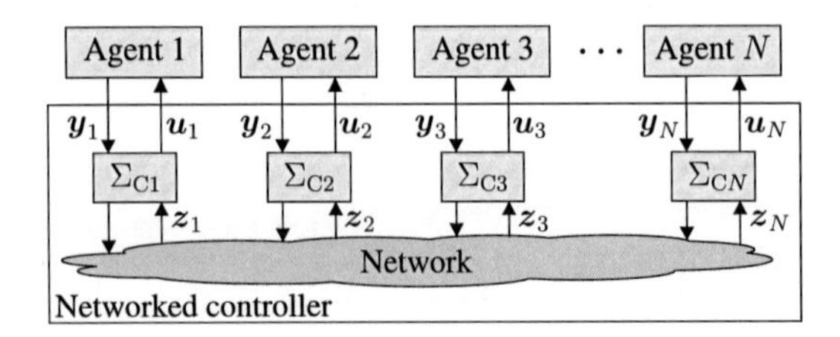

Figure 1.1: Networked controller for the synchronization of multi-agent systems

avoid trivial solutions to the synchronization problem, the agents are assumed to be unstable. The controllers are represented by decentralized units Σ_{Ci}, $(i = 1, 2, \ldots, N)$, which can have any dynamic characteristics. The communication network is used to represent the information exchange between the local controllers, which can either be directed or undirected. Furthermore it is assumed throughout this thesis that the communication network is instantaneous, lossless and connected.

1.2 Synchronization problem and performance specifications

This thesis investigates the synchronization problem of multi-agent systems, where the agents outputs $\boldsymbol{y}_i(t)$ have to satisfy the control goal

Asymptotic synchronization $$\lim_{t\to\infty} \|\boldsymbol{y}_i(t) - \boldsymbol{y}_{\mathrm{s}}(t)\| = 0, \quad i = 1, 2, \ldots, N. \tag{1.1}$$

The common trajectory $\boldsymbol{y}_{\mathrm{s}}(t)$ is referred to as the synchronous trajectory.

Asymptotic synchronization of networked multi-agent systems has been considered in many applications, for example, in the coordination of autonomous agents including formation control, flocking or agent rendezvous problems (see e.g. [24], [25], [26] and the references therein). Some of the earliest references on methods for the design of distributed synchronizing controllers are [27], [28] and [29]. The vast majority of the literature in this area has focused on the asymptotic synchronization of complex agent dynamics with simple communication structures and vice versa. The latter is often considered when investigating consensus problems.

In contrast to the literature, this thesis focuses on methods that do not only fulfill the requirement on asymptotic synchronization but also requirements on the transient behavior of the agents, which are specified as follows.

Optimal synchronization

The networked controller should be designed to minimize the quadratic performance index

$$J(\boldsymbol{x}_i(0)) = \int_0^\infty \sum_{i=1}^{N} \sum_{j=i+1}^{N} |l_{ij}| \left(\boldsymbol{x}_i(t) - \boldsymbol{x}_j(t)\right)^T \boldsymbol{Q} \left(\boldsymbol{x}_i(t) - \boldsymbol{x}_j(t)\right) + \boldsymbol{u}_i^T(t)\, \boldsymbol{R}\, \boldsymbol{u}_i(t)\, \mathrm{d}t \quad (1.2)$$

with symmetric positive definite matrices $\boldsymbol{Q}$, $\boldsymbol{R}$ and the scalars $|l_{ij}| \in \{0, 1\}$. The performance index is chosen to penalize the transient behavior of the networked agents with respect to the synchronization errors $\boldsymbol{e}_{ij}(t) = \boldsymbol{x}_i(t) - \boldsymbol{x}_j(t)$.

Synchronization time

The agents should be synchronized within the required five-percent settling time $T_{5\%}$:

$$\left(\sum_{i=1}^{N} \|\boldsymbol{y}_i(t) - \boldsymbol{y}_s(t)\|^2\right)^{\frac{1}{2}} \leq \zeta_0, \quad \text{for } t \geq T_{5\%}. \quad (1.3)$$

The scalar ζ_0 is specified in dependence upon the initial conditions $\boldsymbol{x}_i(0)$, $(i = 1, 2, \ldots, N)$.

Damping

The transient behavior of the overall system should satisfy a given minimum damping. For example, a tuning parameter is introduced which is suitable to adjust oscillations in the transient behavior of the overall system.

Note that not all of these requirements have to be fulfilled by each of the developed synchronizing controller design methods. According to the locally available information or the specifications of the communication network, it is useful to focus only on some selected performance requirements.

In this context, it is also important to specify the signals, which are available at each of the local controllers $\Sigma_{\mathrm{C}i}$, $(i = 1, 2, \ldots, N)$. In literature, various techniques have been proposed for couplings to achieve synchronization of multi-agent systems. The most commonly used form are diffusive couplings, where the control input of each agent is specified by a weighted linear combination of output or state signal differences (cf. [30], [31] and [32]). In this thesis diffusive couplings of the form

$$\boldsymbol{z}_i(t) = \sum_{i=1}^{N} l_{ij} \left(\boldsymbol{y}_j(t) - \boldsymbol{y}_i(t)\right) \quad (1.4)$$

with $l_{ij} \in \{-1, 0\}$, $(i \neq j = 1, 2, \ldots, N)$ are considered. That is, the local controller Σ_{Ci} is a feedback of $\boldsymbol{z}_i(t)$ towards $\boldsymbol{u}_i(t)$ as illustrated in Fig. 1.1.

1.3 Literature review

In the 17th century, Christiaan Huygens addressed the synchronization problem after the observation of two clocks swinging in exactly the same frequency. The reason for the synchronized behavior of the clocks was the interaction resulting by small movements of a common frame (cf. [33]). Since then, the significance of couplings among synchronized systems has been subsequently studied for biological and chemical systems and also in physics, where mainly the synchronization of nonlinear or chaotic oscillators has been considered (cf. [34–36]).

Although many technological application examples for synchronization exists, first publications with emphasis on engineering systems arise under the study of averaging problems considered in computer science in the field on distributed computing (see [26] and the references therein). Consensus problems in which the synchronization of integrator systems is considered, were first investigated in [37–40]. The analysis for achieving an agreement regarding a certain quantity of interest of simple agent dynamics that are connected over complex network structures is characteristic for consensus problems. Typically, conditions on the network structure are investigated that have to be satisfied for reaching consensus under a static, a time varying or a delayed information exchange (e.g. [41–43]).

In the synchronization literature, the focus has been laid on the design of distributed controllers for agents with complex dynamics rather than on complex network structures. The design of decentralized controllers was performed for identical agents that start in different initial states $\boldsymbol{x}_i(0)$ and are required to follow a common trajectory $\boldsymbol{y}_{\mathrm{s}}(t)$ (e.g. [44–49]). This trajectory is often chosen to represent the mean behavior, which is given by the autonomous solution of a single agent model with respect to the initial value

$$\boldsymbol{x}_{\mathrm{s}0} = \frac{1}{N} \sum_{i=1}^{N} \boldsymbol{x}_i(0).$$

If the agent network is assumed to have directed communication links, typically the leader-follower synchronization problem is investigated, where the synchronous trajectory $\boldsymbol{y}_{\mathrm{s}}(t)$ is specified by a reference agent known as the leader (eg. [50–52]).

A fundamental aim of the analysis of networked agents is to understand under which conditions the agents can be asymptotically synchronized. The theoretical framework for this synchronizability analysis of linear time-invariant agents with identical and non-identical dynamics was introduced in [53–55] building on the earlier results presented in [56–58]. It was shown by the authors that agents with identical dynamics are synchronizable if and only if a single agent is stabilizable by a static output feedback. In the case of agents with individual dynamics the agents have to posses an internal model of the synchronous trajectory $y_s(t)$ to fulfill eqn. (1.1).

Design methods for distributed controllers have been first developed for agents with identical dynamics and then extended to deal with linear non-identical agents and with nonlinear agents (e.g. [59–64]). Simultaneously, a considerable body of literature has been dedicated to the synchronization analysis of networked agents and the design of synchronizing controllers with respect to real-world networks with time-varying or delayed communication links (e.g. [41, 44, 65, 66]). Further investigations have been made concerning event-based communication, where information between neighboring agents is exchanged over the communication network only if the states of the agents exceed tolerance bounds which are specified by the difference between some locally estimated values and the current state of each agent (eg. [67–72]).

A very important aspect of control engineering concerns the design of controllers to satisfy requirements on the transient behavior of the system to be controlled. Since this aspect has rarely been considered in the literature on synchronization, new design methods are presented in this thesis which address the performance of networked multi-agent systems. The literature which is relevant in relation to the main results of this thesis (presented in Chapters 3 – 5) is reviewed in the following three paragraphs.

LQR and LMI based methods for the design of synchronizing controllers

In [45, 47, 48, 73–77] design methods based on the linear quadratic regulator theory and linear matrix inequalities (LMI) were used to solve the consensus or the synchronization problem for linear multi-agent systems. A computationally expensive solution to a optimization problem for the formation control of discrete-time systems was presented in [78]. It was shown that by separating the controller synthesis problem into an estimation problem and the usual linear quadratic regulator (LQR) control problem a suboptimal solution can be derived which significantly reduces the computational load.

In [73, 79–81] optimal control strategies for single and double-integrator agents (consensus problems) were derived, which are mostly based on the minimization of individual cost functions belonging to each of the agents. The influence of the communication network on the overall system behavior was analyzed in [77]. For a communication network consisting of groups of agents, which are not connected by communication links, it was shown that a full connection of these groups improves the overall system performance.

The analysis of the performance of networked multi-agent systems and the design of a centralized optimal synchronizing controller, which is based on LQR control methods was considered in [76]. The representation of the performance in terms of the L_2 system norm reveals that the transient behavior of the agents is primarily influenced by the second smallest eigenvalue of the coupling matrix $\boldsymbol{L}$. The design of optimal networked controllers for a communication network with a ring and a complete circulant topology was introduced in [82] and [1], respectively. Further investigations on the performance of consensus networks have been made in [83], where the influence of cycles on the $\mathcal{H}_2$ performance was analyzed.

In [47, 49, 74, 84–87] synchronizing networked controllers were designed by solving only one algebraic Riccati equation of a single subsystem order. The corresponding approach was obtained by exploiting the robustness property described by the amplification of the optimal feedback matrix, which only ensures asymptotic synchronization of the agents. In some of these papers, the notion of inverse optimality was used to show that the designed synchronizing controller is optimal with respect to some quadratic performance index. Hence, the corresponding performance index is a result of the design procedure and not of a requirement.

In contrast to the majority of the literature, the methods presented in Chapter 3 address the transient behavior of the networked multi-agent system which is expressed by the requirement of minimizing the quadratic performance index (1.2). The essential difficulty is that the solution of an optimization problem is generally centralized and hence requires an information exchange between all agents. In order to design a networked controller that is optimal in the sense of minimizing the objective function (1.2), a structural constraint is introduced and a gradient-based algorithm is derived that ensures optimal synchronization for a given initial controller matrix $\boldsymbol{K}$, which only fulfills the requirement on asymptotic synchronization.

Output synchronization of multi-agent systems

Asymptotic synchronization by means of a static feedback of the form

$$\boldsymbol{u}_i(t) = -\boldsymbol{K}\,\boldsymbol{z}_i(t) \tag{1.5}$$

with

$$\boldsymbol{z}_i(t) = \sum_{i=1}^{N} l_{ij}\left(\boldsymbol{x}_j(t) - \boldsymbol{x}_i(t)\right) \tag{1.6}$$

has been widely studied in the literature. Since in practice the complete state information is typically not available, recent literature has been focused on methods for the design of networked controllers based on the feedback of diffusive couplings of measured output signals as given in eqn. (1.4). This problem was often solved by a combination of dynamic observers, which are designed to estimate the state information of the diffusive coupling (1.4), and a static feedback in which the coupling signal (1.6) was replaced by its estimate (eg. [46, 47, 62, 88]). In this context, synchronization is achieved under the requirements of observability and stabilizability of a single agent. Observer-based methods for the output synchronization of heterogeneous agents have been presented in [29, 53, 60, 89, 90].

In [89] and [90] homogenization methods were used to align the heterogeneous agent dynamics in order to have homogeneous agents, for which the solution is well known. In [53] and [60] each heterogeneous agent was extended by identical reference models referred to as exosystems, which are tracked by the agents using the exchange of their full state information. Design methods for dynamic synchronizing controllers have e.g. been studied in [91] and [92]. In [93–95] the design of static synchronizing output feedback controllers was considered, where typically LMI-based or optimal control methods are used for the synchronization of agents with identical dynamics. Output synchronization of nonlinear agents was considered for example in [64] and [96].

While earlier publications focus on the asymptotic synchronization of multi-agent systems, recent attempts have been made to also consider the performance of the synchronization process. Both, $\mathcal{H}_2$ and $\mathcal{H}_\infty$ design procedures were used in [97]. The disturbance attenuation of synchronized multi-agent systems was measured in terms of an $\mathcal{H}_\infty$-norm in [98]. Optimal control methods were investigated in [29], where the cooperative behavior of the multi-agent system is improved in transient phases by ensuring a desired decay rate of the synchronization error.

The results presented in Chapter 4 differ from the mentioned design methods in two aspects. The synchronization problem is solved by the design of a **dynamic networked** controller by not only satisfying the requirement on asymptotic synchronization, but also by considering the **transient behavior** with respect to the requirement on the synchronization time and damping. Furthermore, the design method provides a deep insight into the selection procedure of the dynamics of the local controllers.

Synchronization of agent networks with time-varying couplings

As mentioned above, the existing literature on synchronization has mainly focused on the design of synchronizing controllers for static and dynamic networks, but not on networks with coupling restrictions, where each agent is e.g. not able to share information with more than one of its neighbors at the same time.

However, there is an increasing amount of literature considering time-varying networks. In [44] networked controllers for the synchronization of uniformly connected and neutrally stable agents were designed. Synchronization of unstable agents in time-varying networks was considered in [99–101]. Design methods that are related to time varying agent couplings are motivated by real networks, where robustness against link failures or packet losses is of interest. In contrast to the majority of literature on synchronization, the results of Chapter 5 are motivated by the synchronization of multi-agent systems that are connected over peer-to-peer or ad-hoc networks, in which only a pairwise information exchange between agents is allowed. The networked controller has, therefore, to be designed to change the coupling between two agent pairs in a way that synchronization of the overall system can be achieved. Algorithms that ensure average consensus in such restricted networks have received increasing attention in the literature under the notation of gossip algorithms.

It is useful to distinguish between algorithms with a probabilistic agent coupling presented in [102] and a deterministic one considered in [103–105]. In the case of randomized gossip algorithms, the convergence of the agents is considered in a probabilistic sense by studying the expected behavior of the state signals.

In [106] it was shown that a periodic information exchange between agents can be used to achieve a deterministic behavior. In this case, the analysis of the agent behavior can be considered similarly to the consensus problem outlined in [37]. It is worth mentioning that averaging can be achieved by gossiping if and only if the magnitude of the second largest eigenvalue of an appropriate doubly stochastic matrix representing the network is smaller than one. Furthermore, the second smallest eigenvalue of the doubly stochastic matrix quantifies the speed of convergence of the gossip algorithm.

A self-triggered gossip algorithm, where each agent independently determines the time instants for an information exchange, was presented in [107]. Recent publications like [108–111] investigate the improvement of the convergence rate of the algorithms.

However, gossip algorithms are used for the distributed averaging of scalar values, which is similar to the consensus problem or the asymptotic synchronization of simple integrator dynamics. Chapter 5 extends the existing results to the synchronization of unstable discrete-time agents with arbitrary linear dynamics. Moreover, not only asymptotic synchronization is considered, but also the performance of the transient behavior regarding the requirement on the synchronization time as given in (1.3).

1.4 Structure and contribution of this thesis

This thesis is concerned with the synchronization of linear time-invariant multi-agent systems with identical dynamics that are coupled over undirected communication networks with known structure. The focus is on the development of new controller synthesis methods, which ensure the performance requirements on the overall system as presented in Section 1.2. In this context, it is important to investigate the question of how to specify the performance of networked multi-agent systems in view of practical applications.

Two possible types of performance specifications can be formulated with respect to the transient behavior of control systems. Specifications concerning the transient behavior in some time interval, which are typically reflected by a cost functional that is a function of state and control variables and specifications concerning time instants like the requirement on some settling time or overshoot peak. Influenced by such performance specifications, three requirements for the transient behavior of synchronized agents have been introduced in Section 1.2. To reflect the overall system behavior in terms of a cost functional, an infinite time horizon linear quadratic objective function is chosen, which penalizes the state differences between only neighboring agents. Furthermore, the requirement on a settling time of some overall synchronization error and the damping of oscillations is introduced.

According to the performance specifications, three different design approaches are proposed for the synchronization of networked agents by static state feedback controllers, dynamic output feedback controllers and for the synchronization of agent networks with point-to-point connections. The main results of this thesis are presented in Chapters 3 – 5. This chapters are self-contained and can be read independently of one another and in any order. Necessary preliminaries are presented in Chapter 2, which is organized as follows.

Chapter 2 – Preliminaries

The first part of this chapter introduces basic mathematical notations and some important definitions which are used throughout this thesis. The second part presents well-known results from graph theory that are needed for the modeling of the communication network. The LQR design approach and the root locus design approach are reviewed in the third part of this chapter. These two approaches serve as a basis for the design methods presented in Chapters 3 and 4, respectively. According to Chapter 5, the basic idea of gossip algorithms is presented in Section 2.5. Finally, a plant for the synchronization of autonomous mobile robots is described which is used as an example in order to illustrate the quality of the proposed controller design methods.

Chapter 3 – Optimal synchronization

The first part of this thesis presents LQR methods for the synchronization of continuous-time agents that are networked by static controllers $\boldsymbol{K}$ of the form (1.5) – (1.6). Based on a decomposition of the overall system model, a necessary and sufficient condition is presented in Theorem 3.1 under which the agents are synchronized.

Section 3.6 introduces an LQR problem for the design of the synchronizing networked controllers. It is shown that a decomposition of the optimization problem is necessary to exclude the synchronous behavior in order to obtain the optimal solution. This decomposition is applied to the LQR optimization problem for synchronization, where the minimization of (1.2) with respect to the control input of the agents is considered (Lemma 3.1). The solution of the decomposed optimization problem is presented in Lemma 3.2 and the corresponding optimal controller in Theorem 3.2. The solution of the LQR problem leads to a centralized controller which ensures *optimal synchronization of completely coupled agents.*

A design method for *asymptotic synchronization of agents coupled over non-complete networks* is investigated in Section 3.8. This design method uses an approximation approach developed in Theorem 3.3 to minimize the deviation between the optimal networked controller provided by Theorem 3.2 and the desired structure of a networked controller, which is given by the separation into identical local controllers and the communication network.

The main contribution of this chapter consists in a design method for *optimal synchronization of non-complete agent networks*. The idea for the design method is based on a modified optimization problem, which considers the structure of the communication network as an additional constraint (Section 3.9). The solution of this extended optimization problem is obtained in Lemma 3.3 and Theorem 3.4, where it is shown that the optimization problem is computationally expensive to solve. For this reason a gradient descent algorithm is proposed that ensures optimal synchronization for any given local controllers $\boldsymbol{K}$ satisfying the weak requirement on asymptotic synchronization (Algorithm 3.1).

Section 3.10 illustrates the design methods by its experimental application to the control of a vehicle platoon.

Chapter 4 – Output synchronization

The second part of this thesis presents an algorithm for the design of dynamic controllers, which solves the output synchronization problem of single-input/single-output (SISO) agents that are affected by disturbances as shown in Fig. 1.2. Inspired by the root locus design method, the control algorithm is designed to ensure the requirements on the synchronization time and a minimum damping of the networked multi-agent system for arbitrary undirected and most of directed communication networks. The design of the dynamic controllers Σ_C

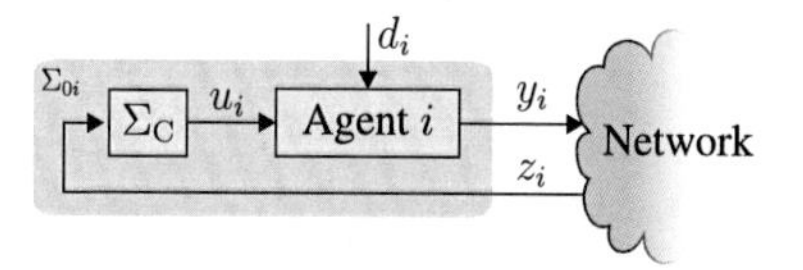

Figure 1.2: Networked SISO agent with disturbance

is based on the analysis of how the network topology influences the behavior of some decomposed error dynamics derived in Lemma 4.1 and Theorem 4.1. This analysis reveals the following two interesting properties:

1. The dynamics of the decomposed synchronization errors are given by $N-1$ decoupled closed-loop systems.

2. The influence of the network on the decomposed synchronization errors is similar to the influence of a static feedback on the $N-1$ open loops Σ_{0i}, $(i = 2, 3, \ldots, N)$. In detail, the error dynamics are obtained by closing the open loops with the corresponding static gain λ_i, which is specified by the network: $z_i(t) = \lambda_i \, y_i(t)$.

The key assumption for the design of the controllers Σ_C is based on the creation of a dominant pair of poles in the dynamics of the decomposed synchronization errors. This procedure leads to a transient behavior of the decoupled synchronization errors, which is bounded with respect to the second smallest eigenvalue λ_2 of the Laplacian matrix of the communication network (Lemma 4.3). The eigenvalue λ_2 is known in the literature as the algebraic connectivity representing a measure of the convergence rate of consensus algorithms (cf. [26] and [112]). The analysis in Chapter 4 shows that the algebraic connectivity has a significant influence on the behavior of the overall system even when considering agents with general dynamics.

The main idea of the controller design procedure is to bring the bound of the decomposed synchronization errors into a relation with the synchronization error of the overall system (Theorem 4.2) and to translate the requirements on the synchronization time and the minimum damping into specifications of the upper bound, which are afterwards used for the determination of the controller parameters (Section 4.6.4 and Algorithm 4.1). Finally, the design approach is extended by the use of homogenization methods for the synchronization of multi-agent systems with non-identical dynamics in Section 4.7.1 and directed communication networks in Section 4.7.2.

The results are demonstrated by its experimental application to the synchronization of a vehicle platoon in Section 4.8.

Chapter 5 – Synchronization in point-to-point networks

The third part proposes two decentralized control algorithms for the synchronization of identical discrete-time agents that are connected over a communication network, where each agent can exchange information with only one of its neighbors at the same time. In order to achieve synchronization in this type of networks, the interconnection between the agents have to be considered as a part of the design procedure.

The underlying idea of the proposed approaches is to split up the local control units Σ_{Ci} shown in Fig. 1.1 into two components. The first component is referred to as the communication unit and the second component is a state feedback controller. Assume that the communication unit is designed to establish an exchange of the state information between two neighboring agents that are currently not connected to any other agent in a deterministic or a probabilistic repetitive order. Then, asymptotic synchronization of the multi-agent system can be achieved if the state feedback controller is designed to ensure complete synchronization of any coupled agent pair in a finite time interval that is available for the coupling. The difference between the two control algorithms consists in the choice of the neighboring agents, which is either periodically repeating or random (Section 5.4 and 5.5). The synchronization of a coupled pair of agents in finite time is achieved by the use of a dead-beat controller, which ensures synchronization in at most n time steps, where n is the dynamic order of the agents.

The main contribution of this part is the synchronization analysis for agents coupled by control algorithms with a deterministic or a random coupling sequence and the corresponding necessary and sufficient conditions for asymptotic synchronization derived in Theorem 5.1 and 5.3. This conditions reveal that asymptotic synchronization of unstable agents can only be achieved if the complete synchronization of the coupled agent pairs appears faster than the divergence of their unstable modes. Since a dead-beat controller is used for the synchronization of the coupled agents in finite time, it is obvious that a reduction of the sampling time of the discrete-time agents directly reduces the time after which the coupled agents are synchronized. Hence, asymptotic synchronization can be achieved by adjusting the sampling time. Furthermore, reducing the sampling time directly reduces the synchronization time of the networked agents. This relation is used for the design of the sampling time to satisfy the requirement on the synchronization time in Theorem 5.2 and 5.4.

Section 5.6 illustrates the control algorithms by its application to the synchronization of harmonic oscillators.

Chapter 6

This chapter summarizes the results of this thesis.

2 Preliminaries

2.1 Notations and definitions

Throughout this thesis, scalars are represented by italic letters ($a \in \mathbb{R}$), vectors by bold face letters and matrices by upper-case bold face letters ($\boldsymbol{x} \in \mathbb{R}^n$, $\boldsymbol{X} \in \mathbb{R}^{n\times n}$). The entire set of real numbers is denoted by $\mathbb{R}$. The set of natural numbers is denoted by $\mathbb{N}$ and the set of integers by $\mathbb{Z}$. $\boldsymbol{I}$ represents the identity matrix of appropriate dimension. The i-th element of the j-th column of the matrix $\boldsymbol{A} \in \mathbb{R}^{n\times m}$ is given by $(\boldsymbol{A})_{ij}$ and the i-th entry of a vector $\boldsymbol{x}$ by $(\boldsymbol{x})_i$, respectively. $\mathbb{1} = (\,1\,\dots\,1\,)^T$ is the one vector, $\mathbf{0} = (\,0\,\dots\,0\,)^T$ the zero vector. $\Re\{z\}$ denotes the real part of a complex number z and $\Im\{z\}$ the imaginary part. The matrix $\boldsymbol{M} \in \mathbb{R}^{\mathrm{r}\times\mathrm{c}}$ with the entries m_{ij}, $i = 1, 2, \dots, \mathrm{r}$ and $j = 1, 2, \dots, \mathrm{c}$ is denoted by $\boldsymbol{M} = (m_{ij})$.

The absolute value of the scalar a is symbolized by $|a|$. Sets are denoted by calligraphic letters ($\mathcal{A}$, $\mathcal{P}$) and their cardinality or size by $|\mathcal{P}|$. The expression $\boldsymbol{A} \succ 0$ describes the positive definiteness of the matrix $\boldsymbol{A}$. The trace of a square matrix $\boldsymbol{A} \in \mathbb{R}^{n\times n}$ is symbolized by $\mathrm{tr}\,(\boldsymbol{A})$. Similarly, the rank of a matrix $\boldsymbol{A} \in \mathbb{R}^{n\times m}$ is denoted by $\mathrm{rank}\,(\boldsymbol{A})$. The i-th eigenvalue of $\boldsymbol{A} \in \mathbb{R}^{n\times n}$ is denoted by $\lambda_i\,(\boldsymbol{A})$ and its spectral radius by

$$\rho(\boldsymbol{A}) = \max\left\{|\lambda_i\,(\boldsymbol{A})\,|,\ i = 1, 2, \dots, n\right\}.$$

The largest eigenvalue of a matrix $\boldsymbol{A} \in \mathbb{R}^{n\times n}$ is defined as

$$\lambda_{\max}\,(\boldsymbol{A}) = \max\left\{\lambda_i\,(\boldsymbol{A}),\ i = 1, 2, \dots, n\right\}.$$

$\mathrm{diag}(\lambda_i)$ is a diagonal matrix with the diagonal entries $\lambda_1, \lambda_2, \dots, \lambda_N$ and

$$\mathrm{diag}\,(\boldsymbol{A}_i) = \begin{pmatrix} \boldsymbol{A}_1 & 0 & \cdots & 0 \\ 0 & \boldsymbol{A}_2 & \ddots & \vdots \\ \vdots & \ddots & \ddots & 0 \\ 0 & \cdots & 0 & \boldsymbol{A}_N \end{pmatrix}$$

a block diagonal matrix, respectively.

The step signal is defined as

$$\sigma(t) = \begin{cases} 1, & t \geq 0 \\ 0, & \text{otherwise.} \end{cases}$$

The ceiling function

$$\lceil x \rceil = \min\{n \in \mathbb{Z} \,|\, n \geq x\}$$

gives the smallest integer that is larger than or equal to x.

Transformations into the frequency domain are symbolized by $F(s) \bullet\!\!-\!\!\circ f(t)$ with the Laplace transform $F(s)$ of the continuous-time signal $f(t)$. The expected value of a discrete random variable $X \in \{x_1, x_2, \ldots, x_N\}$ with the associated probability distribution $f(x_i) = p_i$ is defined by

$$\mathrm{E}(X) = \sum_{i=1}^{N} p_i \, x_i.$$

$\|\boldsymbol{x}\|$ denotes the euclidean vector norm of $\boldsymbol{x} \in \mathbb{R}^n$ and $\|\boldsymbol{A}\|$ the compatible spectral norm of the matrix $\boldsymbol{A} \in \mathbb{R}^{n \times n}$, which are defined by

$$\|\boldsymbol{x}\| = \sqrt{\sum_{i=1}^{N} (\boldsymbol{x})_i^2}$$

and

$$\|\boldsymbol{A}\| = \sqrt{\lambda_{\max}\left(\boldsymbol{A}^T \boldsymbol{A}\right)},$$

respectively. The Frobenius norm of the matrix $\boldsymbol{A} \in \mathbb{R}^{n \times m}$, denoted by $\|\boldsymbol{A}\|_{\mathrm{F}}$, is defined by

$$\|\boldsymbol{A}\|_{\mathrm{F}} = \sqrt{\sum_{i=1}^{n} \sum_{j=1}^{m} (\boldsymbol{A})_{ij}^2}.$$

Finally, $\boldsymbol{A} \otimes \boldsymbol{B}$ denotes the Kronecker product of two matrices $\boldsymbol{A} = (a_{ij}) \in \mathbb{R}^{n \times n}$ and $\boldsymbol{B} \in \mathbb{R}^{m \times m}$ which is defined by

$$\boldsymbol{A} \otimes \boldsymbol{B} = \begin{pmatrix} a_{11}\boldsymbol{B} & a_{12}\boldsymbol{B} & \cdots & a_{1n}\boldsymbol{B} \\ a_{21}\boldsymbol{B} & a_{22}\boldsymbol{B} & \ddots & \vdots \\ \vdots & \ddots & \ddots & a_{(n-1)n}\boldsymbol{B} \\ a_{n1}\boldsymbol{B} & \cdots & a_{n(n-1)}\boldsymbol{B} & a_{nn}\boldsymbol{B} \end{pmatrix}. \tag{2.1}$$

From (2.1) it is easy to see that

$$\operatorname{diag}(\lambda_i) \otimes \boldsymbol{A} = \operatorname{diag}(\lambda_i \boldsymbol{A}).$$

See [113] and the references therein for an overview on Kronecker algebra related to system theory. The following lemmas present some important properties of the Kronecker product.

Lemma 2.1 (Theorem 13.3, 13.4 and 13.6 in [114]). *Let $\boldsymbol{A} \in \mathbb{R}^{r\times n}$, $\boldsymbol{B} \in \mathbb{R}^{r\times s}$, $\boldsymbol{C} \in \mathbb{R}^{n\times p}$, $\boldsymbol{D} \in \mathbb{R}^{s\times t}$, $\boldsymbol{E} \in \mathbb{R}^{r\times r}$ with $\operatorname{rank}(\boldsymbol{E}) = r$ and $\boldsymbol{F} \in \mathbb{R}^{n\times n}$ with $\operatorname{rank}(\boldsymbol{F}) = n$. Then*

$$\begin{aligned} (\boldsymbol{A}\otimes\boldsymbol{B})(\boldsymbol{C}\otimes\boldsymbol{D}) &= (\boldsymbol{AC}\otimes\boldsymbol{BD}) \\ (\boldsymbol{A}\otimes\boldsymbol{B})^T &= \left(\boldsymbol{A}^T\otimes\boldsymbol{B}^T\right) \\ (\boldsymbol{E}\otimes\boldsymbol{F})^{-1} &= \left(\boldsymbol{E}^{-1}\otimes\boldsymbol{F}^{-1}\right). \end{aligned}$$

Lemma 2.2 (First Theorem of Sec. 2.6 in [115]). *Let $\boldsymbol{A} \in \mathbb{R}^{n\times n}$ with eigenvalues λ_i, $(i = 1, 2, \dots, n)$ and $\boldsymbol{B} \in \mathbb{R}^{m\times m}$ with eigenvalues μ_i, $(i = 1, 2, \dots, m)$. Then the largest magnitude of the eigenvalues of $\boldsymbol{A}\otimes\boldsymbol{B}$ is given by*

$$\rho(\boldsymbol{A}\otimes\boldsymbol{B}) = \rho(\boldsymbol{A})\,\rho(\boldsymbol{B}).$$

Lemma 2.3. *Let $\boldsymbol{A} \in \mathbb{R}^{n\times n}$ and $\boldsymbol{B} \in \mathbb{R}^{m\times m}$. Then the spectral norm of the Kronecker product $\boldsymbol{A}\otimes\boldsymbol{B}$ is given by*

$$\|\boldsymbol{A}\otimes\boldsymbol{B}\|^2 = \|\boldsymbol{A}\|^2\,\|\boldsymbol{B}\|^2.$$

Proof. In view of Lemma 2.1 and Lemma 2.2, the proof is obtained by

$$\begin{aligned} \|\boldsymbol{A}\otimes\boldsymbol{B}\|^2 &= \rho\left((\boldsymbol{A}\otimes\boldsymbol{B})^T(\boldsymbol{A}\otimes\boldsymbol{B})\right) \\ &= \rho\left(\boldsymbol{A}^T\boldsymbol{A}\otimes\boldsymbol{B}^T\boldsymbol{B}\right) \\ &= \rho\left(\boldsymbol{A}^T\boldsymbol{A}\right)\,\rho\left(\boldsymbol{B}^T\boldsymbol{B}\right) \\ &= \|\boldsymbol{A}\|^2\,\|\boldsymbol{B}\|^2. \end{aligned}$$

□

2.2 Graph theory

General properties and definitions

Most of the methods presented in this thesis assume undirected graphs that are used to describe the structure of the communication network. An undirected graph $\mathcal{G} = (\mathcal{V}, \mathcal{E})$ consists of a vertex set $\mathcal{V} = \{v_1, v_2, \ldots, v_N\}$ and an edge set $\mathcal{E} = \{\boldsymbol{e}_1, \boldsymbol{e}_2, \ldots, \boldsymbol{e}_M\} \subseteq \mathcal{V} \times \mathcal{V}$, where each edge $\boldsymbol{e}_k = \{v_i, v_j\}$, $(k = 1, 2, \ldots, M)$ is defined by a pair of two distinct vertices. The i-th vertex $v_i \in \mathcal{V}$ of the graph $\mathcal{G}$ is associated with the i-th agent and an edge $\boldsymbol{e}_k = \{v_i, v_j\} \in \mathcal{E}$ with a coupling between the i-th and the j-th agent. If $\{v_i, v_j\} \in \mathcal{E}$, then agent j is referred to as a *neighbor* of agent i. The set of neighbors of agent i is defined by

$$\mathcal{N}_i = \{v_j \mid \{v_i, v_j\} \in \mathcal{E}\}.$$

The cardinality of the set of neighbors gives the number of neighboring agents, which will be denoted by $|\mathcal{N}_i|$.

Incidence and Laplacian matrices are used in this thesis to describe the *communication graph* that represents the structure of the network mathematically. The incidence matrix $\boldsymbol{M}$ of the undirected graph $\mathcal{G}$ with N vertices and M edges is a matrix of order $N \times M$ where

$$(\boldsymbol{M})_{ij} = \begin{cases} 1 & \text{if } v_i \text{ is the first entry in } \boldsymbol{e}_j, \\ -1 & \text{if } v_i \text{ is the second entry in } \boldsymbol{e}_j, \\ 0 & \text{otherwise} \end{cases} \tag{2.2}$$

for every $\boldsymbol{e}_j \in \mathcal{E}$, $(j = 1, 2, \ldots, M)$. The elements of the Laplacian matrix $\boldsymbol{L} \in \mathbb{R}^{N \times N}$ of the graph $\mathcal{G}$ are defined by

$$(\boldsymbol{L})_{ij} = \begin{cases} d_i & \text{if } i = j, \\ -1 & \text{if } \{v_i, v_j\} \in \mathcal{E}, \\ 0 & \text{else.} \end{cases} \tag{2.3}$$

$d_i = |\mathcal{N}_i|$ represents the number of edges connected to node v_i. Another important representation of the Laplacian matrix is given by

$$\boldsymbol{L} = \boldsymbol{M}\boldsymbol{M}^T \tag{2.4}$$

$$= \sum_{k=1}^{M} \epsilon_k \epsilon_k^T \tag{2.5}$$

with ϵ_i representing the i-th column of the incidence matrix $\boldsymbol{M} = (\epsilon_1 \; \epsilon_2 \; \cdots \; \epsilon_M)$. In this context, the i-th and j-th agents are said to be coupled if and only if $(\boldsymbol{L})_{ij} \neq 0$. It is well known from the literature that synchronization of unstable agents can only be achieved if each agent is coupled to every other agent over the communication network. In terms of the graph theory, the graph $\mathcal{G}$ has to contain a *spanning tree* (cf. [116]). The following lemma formulates an important relation between the existence of a spanning tree in $\mathcal{G}$ and the second eigenvalue of the corresponding Laplacian matrix.

Lemma 2.4 (cf. Theorem 2.1 in [117]). *Let $\mathcal{G}$ be a graph described by the Laplacian matrix $\boldsymbol{L}$ with eigenvalues $\lambda_1 \leq \lambda_2 \leq \ldots \leq \lambda_N$. Then, the Graph $\mathcal{G}$ has a spanning tree if and only if $\lambda_2 > 0$.*

Hence, the graph $\mathcal{G}$ is referred to as *connected* if $\lambda_2(\boldsymbol{L}) > 0$. Note that the Laplacian $\boldsymbol{L}$ of undirected graphs is symmetric. Consequently, there exists an orthogonal transformation matrix $\boldsymbol{T} = (t_1 \; t_2 \; \cdots \; t_N)$ such that $\boldsymbol{T}^T = \boldsymbol{T}^{-1}$ and

$$\boldsymbol{T}^T \boldsymbol{L} \boldsymbol{T} = \operatorname{diag}(\lambda_i(\boldsymbol{L})) \tag{2.6}$$

holds. Since all row sums of the Laplacian matrix $\boldsymbol{L}$ vanish, the Laplacian always has a zero eigenvalue $\lambda_1(\boldsymbol{L}) = 0$ with the eigenvector

$$\boldsymbol{t}_1 = \frac{1}{\sqrt{N}} \mathbb{1}.$$

Graphs of pairwise connected vertices

This section presents results that are used in Chapter 5. If every agent is allowed to be coupled to only one other agent, then it is obvious that the corresponding Laplacian matrix given by (2.5) fulfills

$$\epsilon_i^T \epsilon_j = 0 \;\; \forall \;\; i \neq j, \;\; i, j = 1, 2, \ldots, M. \tag{2.7}$$

An example for the interconnection of pairwise coupled agents is illustrated in Fig. 2.1.

Consider

$$\mathcal{L} = \left\{ \boldsymbol{L} = \boldsymbol{M}^T \boldsymbol{M} \mid \epsilon_k^T \epsilon_l = 0 \;\; \forall \;\; k \neq l, \;\; k, l = 1, 2, \ldots, M \right\}$$

as the set of all Laplacian matrices (2.5) satisfying property (2.7). For these Laplacian matrices it is easy to verify that

$$\boldsymbol{L}_i^2 = 2 \boldsymbol{L}_i \;\; \forall \;\; \boldsymbol{L}_i \in \mathcal{L} \tag{2.8}$$

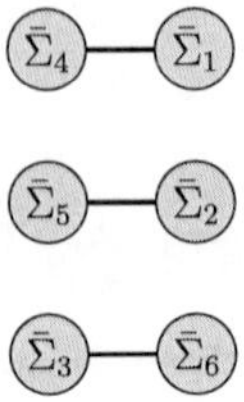

$$\boldsymbol{L} = \begin{pmatrix} 1 & 0 & 0 & -1 & 0 & 0 \\ 0 & 1 & 0 & 0 & -1 & 0 \\ 0 & 0 & 1 & 0 & 0 & -1 \\ -1 & 0 & 0 & 1 & 0 & 0 \\ 0 & -1 & 0 & 0 & 1 & 0 \\ 0 & 0 & -1 & 0 & 0 & 1 \end{pmatrix}$$

Figure 2.1: Pairwise coupled agents

holds. Another important representation of graphs satisfying (2.7) is given by the matrices

$$\boldsymbol{T}_i = \boldsymbol{I} - \frac{1}{2}\boldsymbol{L}_i, \quad \boldsymbol{L}_i \in \mathcal{L}, \tag{2.9}$$

which have some interesting properties. Namely, $\boldsymbol{T}_i \mathbb{1} = \mathbb{1}$ and $\mathbb{1}^T \boldsymbol{T}_i = \mathbb{1}^T$ of the non-negative matrices $\boldsymbol{T}_i$ show that these are doubly stochastic and hence have at least one eigenvalue $\lambda_1(\boldsymbol{T}_i) = 1$ with the eigenvector

$$\boldsymbol{v}_1 = \frac{1}{\sqrt{N}}\mathbb{1}. \tag{2.10}$$

From (2.8) and (2.9) it is easy to see that

$$\boldsymbol{T}_i^n = \boldsymbol{T}_i \quad \text{and} \quad \boldsymbol{T}_i\boldsymbol{L}_i = \boldsymbol{L}_i\boldsymbol{T}_i = \boldsymbol{O}. \tag{2.11}$$

The following lemma presents a matrix product which is often used in the literature on periodic gossip algorithms. The matrix product appears to be important for the synchronization analysis considered in Section 5.4.

Lemma 2.5. *Let* $\boldsymbol{T} \in \mathbb{R}^{N \times N}$ *be the matrix product* $\boldsymbol{T} = \boldsymbol{T}_P \cdots \boldsymbol{T}_2 \boldsymbol{T}_1$ *of matrices* $\boldsymbol{T}_i$, $(i = 1, 2, \ldots, P)$ *defined in* (2.9). *Then*

$$\lim_{k \to \infty} \boldsymbol{T}^k = \frac{1}{N}\mathbb{1}\mathbb{1}^T \tag{2.12}$$

if and only if the corresponding graph $\mathcal{G}$ *described by the Laplacian matrix*

$$\boldsymbol{L} = \boldsymbol{L}_1 + \boldsymbol{L}_2 + \ldots + \boldsymbol{L}_P$$

is connected.

Proof. For a proof, see e.g. [103]. □

In the case of random gossiping the matrix

$$\bar{\boldsymbol{T}} = \boldsymbol{I} - \frac{1}{2}\sum_{i=1}^{W} p_i \, \boldsymbol{L}_i \tag{2.13}$$

with

$$\sum_{i=1}^{W} p_i = 1, \qquad p_i \geq 0$$

is of interest. From the definition (2.13) it is easy to verify that $\bar{\boldsymbol{T}}$ is a symmetric doubly stochastic matrix, which has an eigenvalue $\lambda_1\left(\bar{\boldsymbol{T}}\right) = 1$ with the corresponding eigenvector (2.10). Furthermore, there always exists an orthogonal eigenvector matrix $\boldsymbol{V} = (\boldsymbol{v}_1 \; \boldsymbol{v}_2 \; \ldots \; \boldsymbol{v}_N)$ such that

$$\boldsymbol{V}^T \bar{\boldsymbol{T}} \, \boldsymbol{V} = \mathrm{diag}(\bar{\lambda}_i).$$

In general, the eigenvalues of any doubly stochastic matrix $\boldsymbol{T} \in \mathbb{R}^{n\times n}$ fulfill $\lambda_i\left(\boldsymbol{T}\right) \leq 1$, $(i = 1, \ldots, N)$.

The following lemma presents a result considering the convergence of the matrix defined in (2.13). This matrix is important for the analysis of randomly coupled agents presented in Section 5.5.

Lemma 2.6. *Let* $\boldsymbol{L}_i \in \mathcal{L}$, $(i = 1, 2, \ldots, W)$ *be Laplacian matrices satisfying* (2.7) *and* $\bar{\boldsymbol{T}} \in \mathbb{R}^{N\times N}$ *the corresponding doubly stochastic matrix defined by* (2.13). *Then*

$$\begin{aligned} \lim_{k\to\infty} \bar{\boldsymbol{T}}^k &= \boldsymbol{V} \, \mathrm{diag}\left(\lambda_i^k\left(\bar{\boldsymbol{T}}\right)\right) \boldsymbol{V}^T \\ &= \frac{1}{N} \mathbb{1}\mathbb{1}^T \end{aligned} \tag{2.14}$$

if and only if $\lambda_i\left(\bar{\boldsymbol{T}}\right) < 1$, $(i = 2, 3, \ldots, N)$ *which is true if and only if the graph described by* $\bar{\boldsymbol{L}} = \boldsymbol{L}_1 + \boldsymbol{L}_2 + \ldots + \boldsymbol{L}_W$ *is connected.*

Proof. For a proof, see e.g. [102]. □

For basic properties of doubly stochastic matrices and Laplacian matrices see [116] and [118].

2.3 Review of the LQR design method

The aim of optimal control is to determine control inputs for a dynamic system that satisfy given constraints and at the same time minimize a chosen performance criterion (cf. [119], [120]). In the case of the linear-quadratic regulator (LQR) problem, the design of a controller for a linear dynamic system is considered where the performance is reflected by a quadratic objective function. This section presents the basic results of the LQR design method, which are used to determine optimal synchronizing controllers in Chapter 3.

Given the completely controllable dynamic system

$$\dot{\boldsymbol{x}}(t) = \boldsymbol{A}\,\boldsymbol{x}(t) + \boldsymbol{B}\,\boldsymbol{u}(t), \qquad \boldsymbol{x}(0) = \boldsymbol{x}_0,$$

and the quadratic objective function

$$J(\boldsymbol{u},\, \boldsymbol{x}_0) = \int_0^\infty \boldsymbol{x}^T(t)\,\bar{\boldsymbol{Q}}\,\bar{\boldsymbol{Q}}^T\boldsymbol{x}(t) + \boldsymbol{u}^T(t)\boldsymbol{R}\,\boldsymbol{u}(t)\mathrm{d}t, \tag{2.15}$$

where $\boldsymbol{x}(t) \in \mathbb{R}^n$ is the system state, $\boldsymbol{u}(t) \in \mathbb{R}^m$ the control input, $\boldsymbol{R}$ a symmetric positive definite weighting matrix and $\bar{\boldsymbol{Q}}\,\bar{\boldsymbol{Q}}^T$ a positive semi-definite weighting matrix. Then there exists a unique stabilizing optimal feedback

$$\boldsymbol{u}^*(t) = -\underbrace{\boldsymbol{R}^{-1}\boldsymbol{B}^T\boldsymbol{P}}_{\boldsymbol{K}^*}\boldsymbol{x}(t),$$

which solves the optimization problem

$$\begin{aligned}
\min_{\boldsymbol{u}}\; J(\boldsymbol{u},\, \boldsymbol{x}_0) &= \int_0^\infty \boldsymbol{x}^T(t)\,\bar{\boldsymbol{Q}}\,\bar{\boldsymbol{Q}}^T\boldsymbol{x}(t) + \boldsymbol{u}^T(t)\boldsymbol{R}\,\boldsymbol{u}(t)\mathrm{d}t \\
s.t. \qquad \dot{\boldsymbol{x}}(t) &= \boldsymbol{A}\,\boldsymbol{x}(t) + \boldsymbol{B}\,\boldsymbol{u}(t), \qquad \boldsymbol{x}(0) = \boldsymbol{x}_0
\end{aligned}$$

if and only if all unstable modes of the system (2.3) are observable in the objective function (2.15):

$$\operatorname{rank}\begin{pmatrix} \mu_i\boldsymbol{I} - \boldsymbol{A} \\ \bar{\boldsymbol{Q}}^T \end{pmatrix} = n, \quad \text{for all } \mu_i = \lambda_i\left(\boldsymbol{A}\right),\ (i = 1,\, 2,\, \ldots,\, n) \text{ with } \Re\{\mu_i\} \geq 0. \tag{2.16}$$

The matrix $\boldsymbol{P}$ is the symmetric, positive definite solution of the following algebraic riccati equation

$$A^T P + PA - PBR^{-1}B^T P + \bar{Q}\bar{Q}^T = O.$$

2.4 Review of the root locus design method

The root locus design method (cf. [121, 122]) is a well established frequency-domain technique to design controllers for SISO systems. The method is characterized by its graphical insight on how to manipulate the poles of the closed-loop to guarantee a required damping and settling time of the transient response. Therefore, the root-locus method is well suited for the design of controllers that guarantee performance regarding desired reference behavior or desired disturbance rejection.

Performance with respect to the reference input

To guarantee performance specifications on the reference behavior, consider the closed-loop SISO system with the input-to-output transfer function

$$G_{\mathrm{r}}(s) = \frac{Y(s)}{R(s)} = \frac{G(s)\,K(s)}{1 + G(s)\,K(s)}, \tag{2.17}$$

where $Y(s)$ is the output signal, $R(s)$ the reference input, $K(s) = k\,\hat{K}(s)$ the controller and $G(s)$ the plant (cf. Fig. 2.2). The root locus gives the position of the roots of the characteristic polynomial $P(s) = 1 + k\,\hat{K}(s)\,G(s)$ in the complex plane for $k \geq 0$. The scalar k is the gain and the transfer function $\hat{K}(s)$ is used to represent the dynamics of the controller $K(s)$.

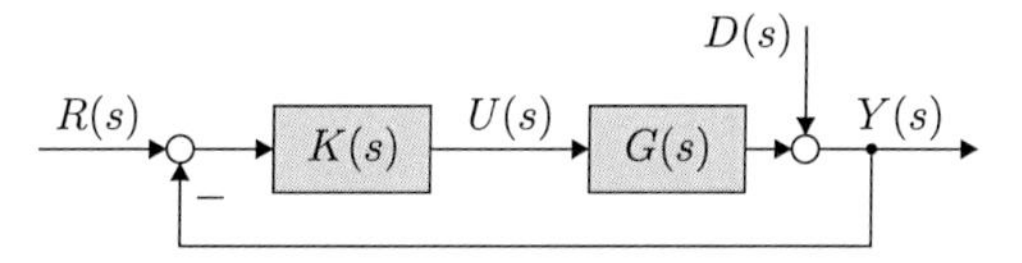

Figure 2.2: Standard closed-loop system

Naturally, the transient response of the transfer function (2.17) shows a complicated form and it is not practicable to guarantee performance by only considering the location of the closed-loop poles in the complex plane. However, it is possible to obtain the transient response if the closed-loop reveals simple dynamics characterized by a dominant pair of poles. In this case, the transfer function (2.17) is approximately described by

$$G_{\mathrm{r}}(s) \approx \frac{1}{\left(\dfrac{s}{\omega_0}\right)^2 + 2\,\dfrac{d}{\omega_0}\,s + 1},$$

where ω_0 is referred to as the natural frequency and d as the damping ratio.

To derive specifications on the transient behavior, the response to a unit step of $G_{\mathrm{r}}(s)$ is considered. It is well known that the step response

$$h_{\mathrm{r}}(t) \circ\!\!-\!\!\bullet\, G_{\mathrm{r}}(s)\frac{1}{s}$$

possesses different time-domain characteristics, which are related to the damping d of the corresponding characteristic polynomial

$$P(s) = \left(\frac{s}{\omega_0}\right)^2 + 2\,\frac{d}{\omega_0}\,s + 1.$$

It has to be distinguished between the underdamped $(d < 1)$, the critically damped $(d = 1)$ and the overdamped case $(d > 1)$ given by

$$h_{\mathrm{r}}(t) = \begin{cases} 1 - \dfrac{\mathrm{e}^{-\omega_0\, d\, t}}{\sqrt{1-d^2}}\,\sin\left(\omega_0\sqrt{1-d^2}\,t + \arccos(d)\right), & d < 1 \\ 1 - (1 + \omega_0\, t)\;\mathrm{e}^{-\omega_0\, t}, & d = 1 \\ 1 - \dfrac{1}{s_1 - s_2}\left(-s_2 \mathrm{e}^{s_1\, t} + s_1 \mathrm{e}^{s_2\, t}\right), & d > 1. \end{cases} \tag{2.18}$$

Note that the poles of the characteristic polynomial $P(s)$ are given by

$$s_{1,2} = -\omega_0\, d \pm \omega_0\sqrt{d^2 - 1}. \tag{2.19}$$

The underdamped case in eqn. (2.18) allows to relate the overshoot Δh and the settling time $T_{5\%}$ of the step response $h_{\mathrm{r}}(t)$ to the parameters d and ω_0 of the system (2.17):

$$\text{Settling time:} \qquad T_{5\%} \approx \frac{3 - \ln\left(\sqrt{1-d^2}\right)}{\omega_0\, d}, \tag{2.20}$$

$$\text{Overshoot peak:} \qquad \Delta h = \mathrm{e}^{-\dfrac{\pi\, d}{\sqrt{1-d^2}}}. \tag{2.21}$$

Hence, the overshoot peak Δh is determined by the damping alone whereas the settling time is obtained from the envelope. Consequently, a given set of performance specifications in terms of the settling time $T_{5\%}$ and the overshoot peak Δh can be translated into specifications on the natural frequency ω_0 and the damping ration d, which are directly related to the location of the dominant pole pair (2.19) in the left hand side of the complex plane.

Performance with respect to the disturbance input

Requirements on the settling time $T_{5\%}$ and the overshoot peak Δh of the closed-loop behavior can also be formulated regarding disturbance inputs. The procedure for the derivation of relations between the performance specifications and the location of the dominant pole pair is similar to the case discussed above.

The disturbance-to-output transfer function

$$\begin{aligned} G_{\mathrm{d}}(s) = \frac{Y(s)}{D(s)} &= \frac{1}{1 + G(s)\,K(s)} \\ &= 1 - G_{\mathrm{r}}(s) && (2.22) \\ &= 1 - \frac{1}{\left(\frac{s}{\omega_0}\right)^2 + 2\frac{d}{\omega_0}s + 1} && (2.23) \end{aligned}$$

shows that the influence of the disturbance on the output of the system is closely related to the input-output behavior of the transfer function $G_{\mathrm{r}}(s)$. In view of eqn. (2.22), the response $h_{\mathrm{d}}(t)$ to a step disturbance at the output of the plant is given by

$$h_{\mathrm{d}}(t) = 1 - h_{\mathrm{r}}(t)$$

where

$$h_{\mathrm{d}}(t) = \begin{cases} \dfrac{\mathrm{e}^{-\omega_0\,d\,t}}{\sqrt{1-d^2}} \sin\left(\omega_0\sqrt{1-d^2}\,t + \arccos(d)\right), & d < 1 \\ (1 + \omega_0\,t)\,\mathrm{e}^{-\omega_0\,t}, & d = 1 \\ \dfrac{1}{s_1 - s_2}\left(-s_2\,\mathrm{e}^{s_1\,t} + s_1\,\mathrm{e}^{s_2\,t}\right), & d > 1 \end{cases} \qquad (2.24)$$

holds.

Based on the given time-domain specifications $(T_{5\%}, \Delta h)$, the controller design is, in both cases, split into two steps. First, the dynamics $\hat{K}(s)$ of the controller have to be chosen such that the closed-loop transfer function has two branches belonging to a dominant pair of poles that cross the specified regions for the locations of the poles. Subsequently the gain k has to be tuned to place the poles inside the required regions (cf. [121, 122]).

2.5 Introduction to gossip algorithms

Gossip algorithms are developed to spread information through multi-agent networks which are, in most cases, engineered to work with limited resources. The main application fields of gossip algorithms are sensor networks, peer-to-peer or wireless ad-hoc networks.

This networks have the property that information can only be exchanged by pairwise communication. In order to achieve a propagation of information through the network, each agent has to exchange his locally available information with all of its neighbors, one after another. Most of the algorithms in the literature use a communication protocol where each agent communicates with a randomly chosen neighbor (cf. [102], [110] and [123]). In contrast to randomized gossip algorithms, there also exist deterministic gossip algorithms in which e.g. each agent communicates with its neighbors according to a periodically repeating sequence (cf. [103] and [106]). In literature there exist a large number of publications on gossiping algorithms from completely different disciplines. In terms of synchronization problems, gossip algorithms for distributed averaging are of interest which can be interpreted as the synchronization of integrator systems.

This section presents the basic idea of random and deterministic gossip algorithms for distributed averaging. Given N integrator systems

$$x_i(k+1) = u_i(k), \quad x_i(0) = x_{0i}$$

and the overall structure of a communication network described by the Laplacian matrix $\boldsymbol{L} = (l_{ij})$. Then the averaging problem given by

$$\lim_{k\to\infty} x_i(k) = \frac{1}{N}\sum_{j=1}^{N} x_{0i}$$

is solved under the gossip constraint if the control input is designed to update the state $x_i(k)$ of the i-th agent and the state $x_j(k)$ of a neighboring agent ($l_{ij} \neq 0$) at time $k+1$ by the average of their values at time k achieved by

$$u_i(k) = u_j(k) = \frac{x_i(k) + x_j(k)}{2}.$$

If there is no available neighbor of agent i at time k, then

$$u_i(k) = 0.$$

An example of a gossip sequence is shown in Fig. 2.3. There are three possible pairwise coupling structures that can appear. For a random selection of a neighboring agent during an

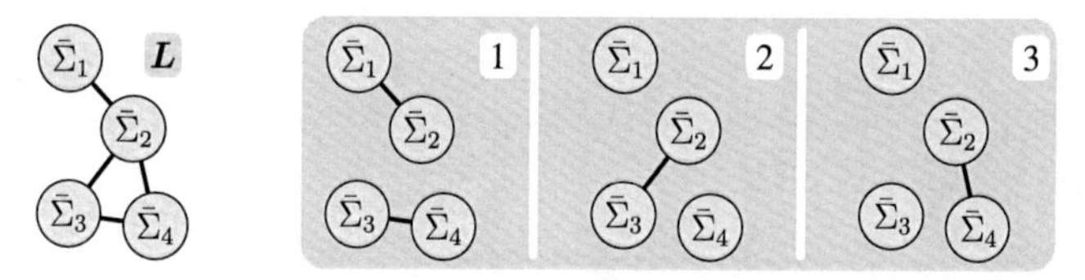

Figure 2.3: Example: Interconnection structure and gossiping sequence

update of the state variable, one of the three coupling structures shown in Fig. 2.3 is chosen with the corresponding probability p_i, $(i = 1, 2, 3)$ at each time instant k. In the case of deterministic gossip, the gossip algorithm is designed to ensure a repetitive coupling of the agents, which is for example achieved by a repetition of the fixed sequence (1-2-3) of the three coupling structures.

Regardless of whether a deterministic or random agent coupling is used, it was shown in the literature that the convergence rate and hence the averaging time of a gossip algorithm depends on the second largest eigenvalue of a doubly stochastic matrix characterizing the algorithm.

2.6 Demonstration example: Autonomous mobile robots

The experimental plant SAMS and the mobile robot e-Puck

The synchronizing controller design methods of Chapter 3 and Chapter 4 are evaluated in experiments by means of a vehicle platoon control realized at the experimental plant ***S**ynchronization of **A**utomnomous **M**obile **S**ystems* (SAMS) at the Institute of Automation and Computer Control at the Ruhr-University Bochum, Germany (Fig. 2.4). The plant was constructed for the simultaneous control of up to seven mobile robots under arbitrary structures of the communication network. The robots used in the plant were designed for education purposes at the École polytechnique fédérale in Lausanne, Switzerland. The so-called e-Pucks are differentially driven mobile robots, where each of the wheels can be independently controlled via a wireless XBee interface by using a personal computer with MATLAB/Simulink.

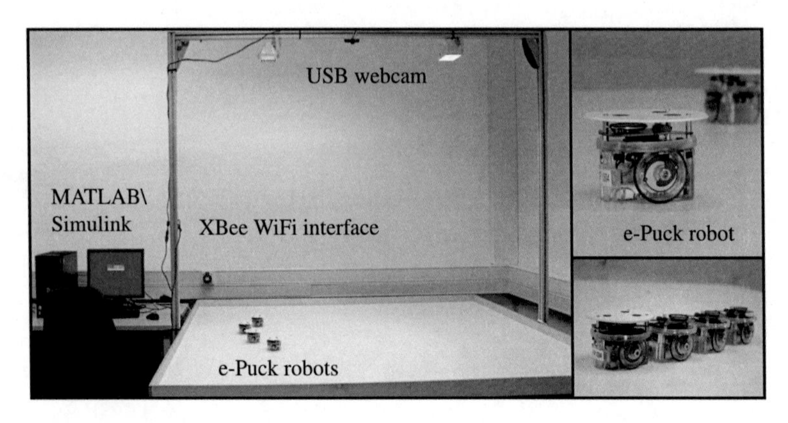

Figure 2.4: Experimental plant: **S**ynchronization of **A**utomnomous **M**obile **S**ystems (SAMS)

Differentially driven mobile robots exhibit complex nonlinear behavior because of nonholonomic constraints in the dynamics. However, the robots are used in this thesis to illustrate the control of the lateral dynamics of a vehicle platoon. Therefore, an orientation controller is implemented to stabilize the movement of each robot in longitudinal direction. Consequently, the state-space model of the i-th robot can be reduced to a description of the one-dimensional movement among the y-axis, which is then given by the linear integral first-order lag system

$$\Sigma_{\mathrm{S}i}: \begin{cases} \dot{\boldsymbol{x}}_i(t) = \begin{pmatrix} 0 & 1 \\ 0 & -0.625 \end{pmatrix} \boldsymbol{x}_i(t) + \begin{pmatrix} 0 \\ 0.028 \end{pmatrix} u_i(t), \quad \boldsymbol{x}_i(0) = \boldsymbol{x}_{0i} \\ \\ y_i(t) = \begin{pmatrix} 1 & 0 \end{pmatrix} \boldsymbol{x}_i(t) \end{cases} \tag{2.25}$$

with the state $\boldsymbol{x}_i^T(t) = \begin{pmatrix} y_i(t) & \dot{y}_i(t) \end{pmatrix}$ and the position $y_i(t)$ in mm. In principal, a simple integrator gives an adequate representation of the lateral dynamics of a mobile robot. However, a prefilter causes the additional first-order lag dynamics in (2.25), since the behavior of the robots is desired to reflect the behavior of real vehicles as far as possible.

Figure 2.5 shows a formation of four mobile robots $\Sigma_{\mathrm{S}i}$ which is used as an application example throughout this thesis in order to illustrate the properties of presented synchronizing controller design methods. The design aim is to ensure the synchronization of the output signals $y_i(t)$ for different initial positions.

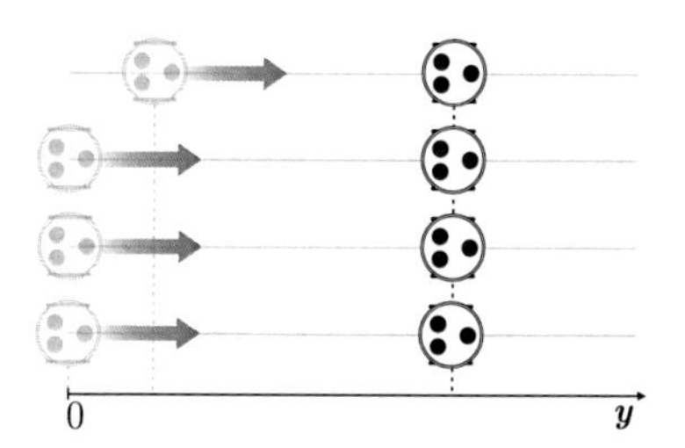

Figure 2.5: Platoon of mobile robots

Basic structure of the experimental plant

As shown in Fig. 2.6, the control scheme of the experimental plant was designed to work in a centralized manner. Thanks to this centralized implementation, arbitrary network structures

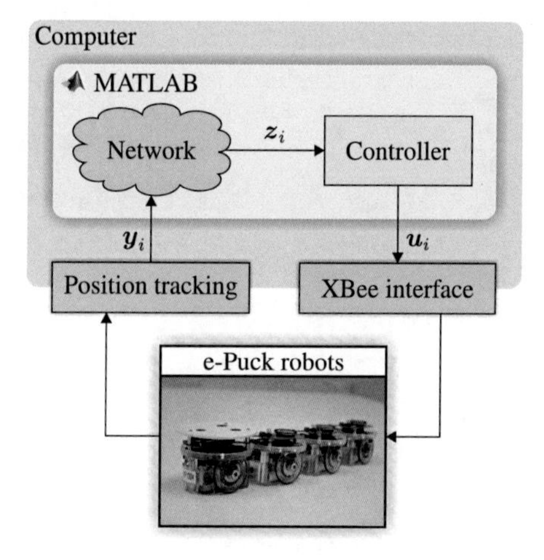

Figure 2.6: Signal flow in the experimental plant SAMS

and local controllers can be emulated using MATLAB.

The tracking of the position and the orientation of the robots is achieved by a measurement software which is activated in the background of the operating systems. The software stores the data automatically in the MATLAB workspace where it can be used for the calculation of the control inputs $u_i(t)$. An XBee interface is used to update the control inputs with the cycle time of $30\,ms$.

3 Optimal synchronization of multi-agent systems

3.1 Introduction to optimal synchronization

This chapter presents methods for the design of static state feedback matrices $\boldsymbol{K}$ to achieve synchronization of N identical unstable agents. The coupling between the agents is realized by the use of a networked controller with the structure shown in Fig. 3.1.

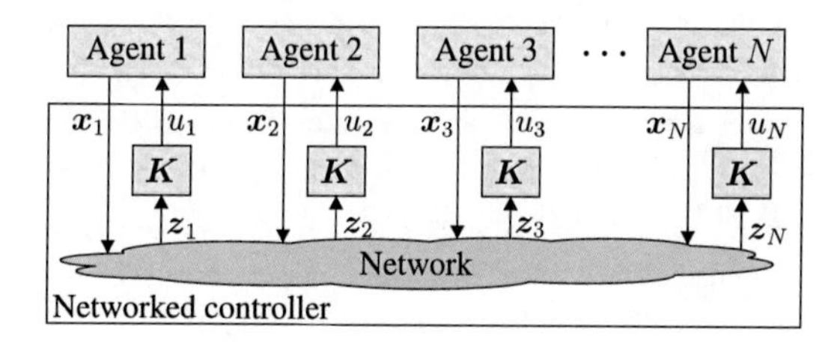

Figure 3.1: State feedback in networked multi-agent systems

As mentioned in the introduction, most methods found in literature solely consider synchronization as an asymptotic property, where the closed-loop system has to satisfy *asymptotic synchronization* which is achieved if

$$\lim_{t\to\infty} \|\boldsymbol{x}_i(t) - \boldsymbol{x}_j(t)\| = 0, \qquad i, j = 1, 2, \ldots, N. \tag{3.1}$$

However, this requirement does not capture the transient behavior of the networked agents which is important, especially when considering application scenarios. Therefore a second requirement is formulated that improves the transient behavior of the agents by requiring a performance index $J(\boldsymbol{K}, \boldsymbol{x}_{0i})$ to be minimized. In this context, let $\mathcal{K}$ be set of controllers fulfilling the requirement on asymptotic synchronization and $\tilde{\mathcal{K}} \subset \mathcal{K}$. A networked multi-agent system fulfills the requirement on *optimal synchronization* if the corresponding local controller $\boldsymbol{K}$ is the solution of an optimization problem

$$\min_{\boldsymbol{K} \in \tilde{\mathcal{K}}} \quad J(\boldsymbol{K},\, \boldsymbol{x}_{0i})$$

with the objective function $J(\boldsymbol{K},\, \boldsymbol{x}_{0i})$ satisfying the following three assumptions

A1) $J(\boldsymbol{K},\, \boldsymbol{x}_{0i}) \to \infty$, for all $\boldsymbol{K} \notin \mathcal{K}$ and $\boldsymbol{x}_{0i} \in \mathbb{R}^n$, $\boldsymbol{x}_{0i} \neq \boldsymbol{x}_{0j}$,

A2) $J(\boldsymbol{K},\, \boldsymbol{x}_{0i}) = 0$, for all $\boldsymbol{x}_{0i} = \tilde{\boldsymbol{x}}_0 \in \mathbb{R}^n$,

A3) $J(\boldsymbol{K},\, \boldsymbol{x}_{0i}) \leq \delta$, for all $\boldsymbol{K} \in \mathcal{K}$ and $\boldsymbol{x}_{0i} \in \mathbb{R}^n$.

To satisfy the requirement on optimal synchronization, a quadratic performance index

$$J(\boldsymbol{K},\, \boldsymbol{x}_{0i}) = \int_0^\infty \sum_{i=1}^{N} \sum_{j=i+1}^{N} |l_{ij}| \left(\boldsymbol{x}_i(t) - \boldsymbol{x}_j(t)\right)^T \boldsymbol{Q} \left(\boldsymbol{x}_i(t) - \boldsymbol{x}_j(t)\right) + \boldsymbol{u}_i^T(t)\, \boldsymbol{R}\, \boldsymbol{u}_i(t)\, \mathrm{d}t. \quad (3.2)$$

is chosen for the design of the networked controller. The matrices $\boldsymbol{Q}$ and $\boldsymbol{R}$ are assumed to be positive definite. The scalars l_{ij} are the elements of the Laplacian matrix $\boldsymbol{L} = (l_{ij})$ representing the communication network and hence the coupling between the agents. The objective function (3.2) reflects the transient behavior of the networked agents regarding the synchronization errors $\boldsymbol{e}_{ij}(t) = \boldsymbol{x}_i(t) - \boldsymbol{x}_j(t)$, $(i,\, j = 1,\, 2,\, \ldots,\, N)$.

The analysis in this chapter shows that asymptotic synchronization of the networked multi-agent system depicted in Fig. 3.1 appears if and only if $N-1$ decoupled error dynamics of a subsystems order are stable. This decoupled error dynamics are significant for designing optimal synchronizing networked controllers, which are based on a decomposition of the objective function.

First, an optimization problem is formulated which is inspired by standard LQR methods. The solution of this optimization problem is derived and it is shown that an all-to-all coupling of the agents is necessary. For networks with a non-complete coupling structure, an approximation method is developed for the design of the local controllers $\boldsymbol{K}$. The approximation is obtained by minimizing the norm of the difference between the centralized optimal feedback and the desired control structure described by the network controller. The approximation method does clearly not minimize the objective function (3.2), but it satisfies the requirement on asymptotic synchronization of the agents for any given communication network containing a spanning tree.

In order to obtain optimal synchronizing controllers, the formulated optimization problem is extended by an additional constraint in the third design method. This constraint is introduced to specify the desired network structure. The resulting optimization problem is non-convex because the networked controller has to satisfy structural constraints. However, a necessary condition for optimality is derived and it is shown that determining the optimal networked controller is computationally expensive. Therefore, an algorithm is presented that uses a gradient method for the design of optimal synchronizing networked controllers.

Structure of this chapter. Section 3.2 introduces the agent model and some basic assumptions. The networked controller and the overall closed-loop system is given in Section 3.3 and Section 3.4, respectively. Based on the overall system model, the synchronization condition is derived in Section 3.5. The design methods that are given in Section 3.7, 3.8 and 3.9 relate on the optimization problem introduced in Section 3.6. Section 3.10 illustrates the design methods by its application to the formation control of mobile robots.

3.2 Agent model

In this chapter a multi-agent system is considered that consists of N identical multi-input multi-output agents described by the state-space models

$$\Sigma_i : \begin{cases} \dot{\boldsymbol{x}}_i(t) = \boldsymbol{A}\,\boldsymbol{x}_i(t) + \boldsymbol{B}\,\boldsymbol{u}_i(t), & \boldsymbol{x}_i(0) = \boldsymbol{x}_{0i} \\ \boldsymbol{y}_i(t) = \boldsymbol{C}\,\boldsymbol{x}_i(t), & i = 1, 2, \ldots, N \end{cases} \tag{3.3}$$

where

- $\boldsymbol{x}_i(t)$ is the n-dimensional state vector,
- $\boldsymbol{u}_i(t)$ the m-dimensional control input and
- $\boldsymbol{y}_i(t)$ the r-dimensional output signal of the i-th agent.

The overall multi-agent system is represented by

$$\Sigma : \begin{cases} \dot{\boldsymbol{x}}(t) = (\boldsymbol{I} \otimes \boldsymbol{A})\,\boldsymbol{x}(t) + (\boldsymbol{I} \otimes \boldsymbol{B})\,\boldsymbol{u}(t), & \boldsymbol{x}(0) = \boldsymbol{x}_0 \\ \boldsymbol{y}(t) = (\boldsymbol{I} \otimes \boldsymbol{C})\,\boldsymbol{x}(t) \end{cases} \tag{3.4}$$

where

$$\begin{aligned} \boldsymbol{x}^T(t) &= \left(\boldsymbol{x}_1^T(t),\, \ldots,\, \boldsymbol{x}_N^T(t)\right), \\ \boldsymbol{u}^T(t) &= \left(\boldsymbol{u}_1^T(t),\, \ldots,\, \boldsymbol{u}_N^T(t)\right), \\ \boldsymbol{y}^T(t) &= \left(\boldsymbol{y}_1^T(t),\, \ldots,\, \boldsymbol{y}_N^T(t)\right) \end{aligned}$$

are vectors and $(\boldsymbol{I} \otimes \boldsymbol{A})$, $(\boldsymbol{I} \otimes \boldsymbol{B})$ and $(\boldsymbol{I} \otimes \boldsymbol{C})$ are block-diagonal matrices with appropriate dimensions.

The requirement (3.1) shows that synchronization of multi-agent systems becomes challenging only when unstable agents are considered. In the case of stable agents, asymptotic synchronization appears without involving a networked controller. Therefore, the following assumption is made.

Assumption 3.1. *The system matrix* $\boldsymbol{A}$ *of the agent models* (3.3) *is assumed to has at least one eigenvalue with a non-negative real part.*

For the application of design methods that are based on optimal control techniques it has furthermore to be assumed that the agents Σ_i, $(i = 1, 2, \ldots, N)$ are completely controllable.

Assumption 3.2. *The agents* Σ_i *are completely controllable*

$$\operatorname{rank}\begin{pmatrix} \boldsymbol{B} & \boldsymbol{A}\boldsymbol{B} & \boldsymbol{A}^2\boldsymbol{B} & \cdots & \boldsymbol{A}^{n-1}\boldsymbol{B} \end{pmatrix} = n.$$

3.3 Static networked controller

The communication network provides the local controllers with the coupling signals $\boldsymbol{z}_i(t)$ that are defined by

$$\boldsymbol{z}_i(t) = \sum_{j=1}^{N} l_{ij}\left(\boldsymbol{x}_j(t) - \boldsymbol{x}_i(t)\right), \qquad i = 1, 2, \ldots, N, \tag{3.5}$$

The factors l_{ij} are the elements of the corresponding Laplacian matrix $\boldsymbol{L} = (l_{ij})$, which can consequently be used to represent the coupling signal $\boldsymbol{z}^T(t) = \left(\boldsymbol{z}_1^T(t),\, \ldots,\, \boldsymbol{z}_N^T(t)\right)$ of the overall system:

$$\boldsymbol{z}(t) = (\boldsymbol{L} \otimes \boldsymbol{I})\, \boldsymbol{x}(t). \tag{3.6}$$

As shown in Fig. 3.1, the networked controller consists of the local static controllers $\boldsymbol{K}$ and the Laplacian matrix $\boldsymbol{L}$ which is associated with the communication network. Note that the structure of the network and hence the corresponding Laplacian matrix $\boldsymbol{L}$, which is also referred to as the communication network $\boldsymbol{L}$, is assumed to be given.

Assumption 3.3. *The communication network described by the Laplacian matrix* $\mathbf{L}$ *is assumed to be undirected and given. Furthermore, the locally available information at each agent is restricted to the coupling signals* (3.5).

This chapter is concerned with the design of the local controllers $\boldsymbol{K}$ as a static feedback

$$\boldsymbol{u}_i(t) = -\boldsymbol{K}\,\boldsymbol{z}_i(t) = \sum_{j=1}^{N} l_{ij}\left(\boldsymbol{x}_j(t) - \boldsymbol{x}_i(t)\right), \qquad i = 1,\, 2,\, \ldots,\, N. \tag{3.7}$$

In terms of the overall system, eqns. (3.6) and (3.7) are combined to obtain the overall control law, which is given by

$$\begin{aligned} \boldsymbol{u}(t) &= -\left(\boldsymbol{I} \otimes \boldsymbol{K}\right)\boldsymbol{z}(t) \\ &= -\left(\boldsymbol{I} \otimes \boldsymbol{K}\right)\left(\boldsymbol{L} \otimes \boldsymbol{I}\right)\boldsymbol{x}(t) \\ &= -\left(\boldsymbol{L} \otimes \boldsymbol{K}\right)\boldsymbol{x}(t). \end{aligned} \tag{3.8}$$

The representation of the control law (3.8) in terms of the Kronecker product involves some benefits, especially for the decomposition of the design problem considered in Section 3.7.1. In virtue of Lemma 2.1, the Kronecker product in the control law (3.8) allows to separate mathematical operations that are related to the network $\boldsymbol{L}$ from those regarding the controller matrix $\boldsymbol{K}$.

3.4 Model of the overall closed-loop system

The state-space model of the overall closed-loop system

$$\begin{aligned} \dot{\boldsymbol{x}}(t) &= \left(\boldsymbol{I} \otimes \boldsymbol{A} - \boldsymbol{L} \otimes \boldsymbol{B}\boldsymbol{K}\right)\boldsymbol{x}(t), \quad \boldsymbol{x}(0) = \boldsymbol{x}_0 \\ \boldsymbol{y}(t) &= \left(\boldsymbol{I} \otimes \boldsymbol{C}\right)\,\boldsymbol{x}(t) \end{aligned} \tag{3.9}$$

is given by the combination of the agent dynamics (3.4) and the feedback (3.8). For the application of optimal control methods, it is necessary to study the state behavior of the overall closed-loop system (3.9) which is obtained by

$$\boldsymbol{x}(t) = \underbrace{\mathrm{e}^{\left(\boldsymbol{I} \otimes \boldsymbol{A} - \boldsymbol{L} \otimes \boldsymbol{B}\boldsymbol{K}\right)t}}_{\boldsymbol{\Phi}(t)}\,\boldsymbol{x}_0. \tag{3.10}$$

3.5 Synchronization condition

The objective of this section is the derivation of a necessary and sufficient synchronization condition for the networked multi-agent system (3.9), which is afterwards used for a specification of requirements regarding the undirected communication network $\boldsymbol{L}$ and the static local controller $\boldsymbol{K}$. Conditions on asymptotic synchronization of networked multi-agent systems were first investigated in [32]. Since then, similar results were obtained in [46, 57, 84] and [124]. However, the following theorem and the corresponding proof is presented for the sake of completeness.

Theorem 3.1 (Synchronization condition). *All agents of the networked multi-agent system* (3.10) *are asymptotically synchronized for all initial conditions* $\boldsymbol{x}_{0i} \in \mathbb{R}^n$, $(i = 1, 2, \ldots, N)$ *if and only if all eigenvalues of the matrices*

$$\tilde{\boldsymbol{A}}_i = \boldsymbol{A} - \lambda_i \boldsymbol{B}\boldsymbol{K}, \quad i = 2, 3, \ldots, N \tag{3.11}$$

have a negative real part.

Proof. See Appendix A.1. □

The theorem shows that asymptotic synchronization is equivalent to the stability of $N-1$ uncoupled closed-loop systems of the order n. Figure 3.2 shows the structure of the closed-

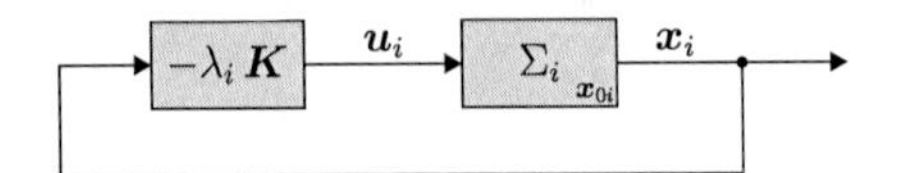

Figure 3.2: Structure of decoupled error dynamics

loop systems that are obtained by the feedback $\boldsymbol{u}_i(t) = -\lambda_i \boldsymbol{K}\boldsymbol{x}_i(t)$, $(i = 2, 3, \ldots, N)$ of the agent dynamics (3.3), which is specified by the eigenvalues λ_i of the Laplacian matrix $\boldsymbol{L}$. Clearly, communication networks with different structures reveal different sets of eigenvalues of corresponding Laplacian matrices. Hence, the aim of designing the networked controller to achieve asymptotic synchronization is equivalent to the design of the controller matrix $\boldsymbol{K}$ to simultaneously stabilize the matrices (3.11) for a given set of eigenvalues λ_i, $(i = 2, 3, \ldots, N)$. According to Assumption 3.3, the considered communication networks are undirected. Thus, the corresponding Laplacian matrices are symmetric and have only real positive eigenvalues $\lambda_1 \leq \lambda_2 \leq \ldots \leq \lambda_N$.

In virtue of Assumption 3.1, Theorem 3.1 can be fulfilled for unstable agents only if $\lambda_2 > 0$. This necessary condition allows to relate the result of Theorem 3.1 to connectivity properties of the communication network. It follows directly from Lemma 2.4 that $\lambda_2 > 0$ if the communication network is connected. In the case of stable agents Theorem 3.1 shows that asymptotic synchronization is achieved even if $\lambda_i = 0$, $(i = 2, 3, \ldots, N)$ holds. Hence, no coupling between the agents is necessary for asymptotic synchronization. The synchronous trajectory of the agents is then represented by $\boldsymbol{y}_{\mathrm{s}}(t) = 0$, which means that the agents asymptotically converge to the origin. However, this chapter considers the non-trivial case where it is assumed that the agents have unstable dynamics.

Corollary 3.1. *In virtue of Assumption 3.1, asymptotic synchronization of the networked multi-agent system* (3.9) *can only be achieved if the communication network is connected.*

Hence, the following assumption is made.

Assumption 3.4. *The communication network described by the Laplacian matrix* $\boldsymbol{L}$ *is assumed to be connected.*

For connected communication networks it was shown in [84] and [47] that there always exists a static controller $\boldsymbol{K}$, which asymptotically synchronizes the agents if and only if all unstable modes of the agent dynamics are controllable by the pair $(\boldsymbol{A}, \boldsymbol{B})$.

3.6 LQR design problem for synchronization

This section introduces an optimization problem for the design of a synchronizing networked controller, where a quadratic objective function is used to represent the performance of the networked agents. Since synchronization is considered, the objective function is chosen to contain terms that exclusively penalize the state differences between neighboring agents. This choice is motivated by applications to autonomous mobile systems. Consider for example a platoon of vehicles or a formation of mobile robots as a control task. In both cases the behavior of neighboring agents is critical, in particular when considering collision avoidance.

The objective function (3.2) defined for synchronization is represented by

$$J(\boldsymbol{u}, \boldsymbol{x}_0) = \int_0^\infty \sum_{\substack{i=1,\\ j=i+1}}^{N} |l_{ij}| \left(\boldsymbol{x}_i(t) - \boldsymbol{x}_j(t)\right)^T \boldsymbol{Q} \left(\boldsymbol{x}_i(t) - \boldsymbol{x}_j(t)\right) + \boldsymbol{u}_i^T(t)\, \boldsymbol{R}\, \boldsymbol{u}_i(t)\, \mathrm{d}t$$

$$= \int_0^\infty \boldsymbol{e}^T(t) \left(\boldsymbol{I} \otimes \boldsymbol{Q}\right) \boldsymbol{e}(t) + \boldsymbol{u}^T(t) \left(\boldsymbol{I} \otimes \boldsymbol{R}\right) \boldsymbol{u}(t)\, \mathrm{d}t \tag{3.12}$$

where $\boldsymbol{Q} = \boldsymbol{Q}^T \succ 0$, $\boldsymbol{R} = \boldsymbol{R}^T \succ 0$ and

$$\boldsymbol{e}(t) = \left(\boldsymbol{M}^T \otimes \boldsymbol{I}\right) \boldsymbol{x}(t). \tag{3.13}$$

Note that $\boldsymbol{M} \in \mathbb{R}^{N \times (|\mathcal{E}|)}$ is the incidence matrix of the communication network defined in (2.2). In the case of a connected communication network, the considered synchronization error $\boldsymbol{e}(t)$ converges asymptotically to zero if and only if the requirement (3.1) on asymptotic synchronization is fulfilled. Consequently the integral in eqn. (3.12) exists for all initial conditions $\boldsymbol{x}_{0i} \in \mathbb{R}^n$, $(i = 1, 2, \ldots, N)$, even if each agent asymptotically follows the unstable synchronous trajectory $\boldsymbol{y}_{\mathrm{s}}(t)$.

According to eqn. (3.13) and (2.4), the weighting of the synchronization error $\boldsymbol{e}(t)$ in the objective function (3.12) can be rewritten in terms of the Laplacian matrix represented by

$$\begin{aligned} \boldsymbol{e}^T(t) \left(\boldsymbol{I} \otimes \boldsymbol{Q}\right) \boldsymbol{e}(t) &= \boldsymbol{x}^T(t) \left(\boldsymbol{M} \otimes \boldsymbol{I}\right) \left(\boldsymbol{I} \otimes \boldsymbol{Q}\right) \left(\boldsymbol{M}^T \otimes \boldsymbol{I}\right) \boldsymbol{x}(t) \\ &= \boldsymbol{x}^T(t) \left(\boldsymbol{M}\boldsymbol{M}^T \otimes \boldsymbol{Q}\right) \boldsymbol{x}(t) \\ &= \boldsymbol{x}^T(t) \left(\boldsymbol{L} \otimes \boldsymbol{Q}\right) \boldsymbol{x}(t). \end{aligned} \tag{3.14}$$

The optimization problem concerned for the design of the networked controller (3.8) is hence given by

$$\min_{\boldsymbol{u}} \; J(\boldsymbol{u}, \boldsymbol{x}_0) = \int_0^\infty \boldsymbol{x}^T(t) \left(\boldsymbol{L} \otimes \boldsymbol{Q}\right) \boldsymbol{x}(t) + \boldsymbol{u}^T(t) \left(\boldsymbol{I} \otimes \boldsymbol{R}\right) \boldsymbol{u}(t)\, \mathrm{d}t \tag{3.15}$$

$$s.t. \qquad \dot{\boldsymbol{x}}(t) = \left(\boldsymbol{I} \otimes \boldsymbol{A}\right) \boldsymbol{x}(t) + \left(\boldsymbol{I} \otimes \boldsymbol{B}\right) \boldsymbol{u}(t), \qquad \boldsymbol{x}(0) = \boldsymbol{x}_0. \tag{3.16}$$

Although the optimization problem (3.15)–(3.16) is very similar to the standard LQR design procedure reviewed in Section 2.3, it can not be directly solved because of the observability requirement in eqn. (2.16), which is not fulfilled when considering synchronization. Since the Laplacian matrix $\boldsymbol{L}$ has a zero eigenvalue ($\lambda_1 = 0$), the weighting matrix $\boldsymbol{L} \otimes \boldsymbol{Q}$ is positive semi-definite even if the matrix $\boldsymbol{Q}$ is positive definite. In particular there exist unstable modes which do not influence the objective function (3.15). That are, if considering identical initial values $\boldsymbol{x}_{0i} = \xi \in \mathbb{R}^n/\{\boldsymbol{0}\}$, $(i = 1, 2, \ldots, N)$ for $\boldsymbol{u}_i(t) = 0$, then it is obvious from (3.12) that $J(\boldsymbol{u}, \boldsymbol{x}_0) = 0$ although the agents follow the unstable synchronous trajectory $\boldsymbol{y}_{\mathrm{s}}(t)$:

$$\boldsymbol{y}_i(t) - \underbrace{\boldsymbol{C}\,\mathrm{e}^{\boldsymbol{A}t}\xi}_{\boldsymbol{y}_{\mathrm{s}}(t)} = \boldsymbol{0}, \qquad i = 1, 2, \ldots, N, \ t \geq 0.$$

This example demonstrates that the objective function (3.15) does obviously not capture the synchronous behavior of the agents. A decomposition of the optimization problem (3.15) – (3.16) is necessary to exclude the synchronous behavior in order to obtain the solution of the optimization problem.

3.7 LQ-Regulator for the synchronization of multi-agent systems

This section deals with the solution of the optimization problem (3.15) – (3.16). A decomposition approach is presented and it is shown that the design of an optimal synchronizing controller can be obtained with respect to modified subsystems that have the order of a single agent. As a consequence, the decomposition reduces the computational complexity of the controller design. However, the solution of the optimization problem (3.2) is presented and it is shown that an all-to-all coupling between the agents is required for optimal synchronization. Hence, the first result of this chapter is a design method for optimally synchronizing controllers of agent networks with a complete coupling structure.

3.7.1 Decomposition of the design problem

The following lemma presents the decomposition of the optimization problem (3.15) – (3.16) which allows to separate the synchronous agent behavior from the behavior of the synchronization error. The optimization problem (3.15) – (3.16) can then be solved by the design of a feedback that ensures the convergence of the synchronization error.

Lemma 3.1 (Decomposition of the optimization problem (3.15)–(3.16)). *Let $\boldsymbol{T}$ be the eigenvector matrix of the Laplacian matrix $\boldsymbol{L}$ with $\boldsymbol{T}^{-1}\boldsymbol{L}\boldsymbol{T} = \operatorname{diag}(\lambda_i)$. Then the decomposition of the optimization problem* (3.15)–(3.16) *is given by the following $i = 1, 2, \ldots, N$ separate optimization problems of order n*

$$\min_{\tilde{\boldsymbol{u}}_i} J_i(\tilde{\boldsymbol{u}}_i, \tilde{\boldsymbol{x}}_{0i}) = \int_0^\infty \lambda_i \tilde{\boldsymbol{x}}_i^T(t)\,\boldsymbol{Q}\,\tilde{\boldsymbol{x}}_i(t) + \tilde{\boldsymbol{u}}_i^T(t)\boldsymbol{R}\,\tilde{\boldsymbol{u}}_i(t)\,\mathrm{d}t \tag{3.17}$$

$$s.t. \qquad \dot{\tilde{\boldsymbol{x}}}_i(t) = \boldsymbol{A}\,\tilde{\boldsymbol{x}}_i(t) + \boldsymbol{B}\,\tilde{\boldsymbol{u}}_i(t), \qquad \tilde{\boldsymbol{x}}_i(0) = \tilde{\boldsymbol{x}}_{0i} \tag{3.18}$$

where

$$J(\boldsymbol{u}, \boldsymbol{x}_0) = \sum_{i=1}^{N} J_i(\tilde{\boldsymbol{u}}_i, \tilde{\boldsymbol{x}}_{0i}),$$

$$\tilde{\boldsymbol{x}}(t) = \left(\boldsymbol{T}^{-1} \otimes \boldsymbol{I}\right) \boldsymbol{x}(t), \tag{3.19}$$

$$\tilde{\boldsymbol{u}}(t) = \left(\boldsymbol{T}^{-1} \otimes \boldsymbol{I}\right) \boldsymbol{u}(t), \tag{3.20}$$

$\tilde{\boldsymbol{x}}^T(t) = \left(\tilde{\boldsymbol{x}}_1^T(t), \ldots, \tilde{\boldsymbol{x}}_N^T(t)\right)$ *and* $\tilde{\boldsymbol{u}}^T(t) = \left(\tilde{\boldsymbol{u}}_1^T(t), \ldots, \tilde{\boldsymbol{u}}_N^T(t)\right)$.

Proof. The proof of Lemma 3.1 is obtained as the result of the state transformation (3.19) and the input transformation (3.20) of the optimization problem (3.15)–(3.16). The state transformation (3.19) of the weighting term (3.14) gives

$$\begin{aligned}
\boldsymbol{x}^T(t)\,(\boldsymbol{L} \otimes \boldsymbol{Q})\,\boldsymbol{x}(t) &= \boldsymbol{x}^T(t)\left(\boldsymbol{T}\boldsymbol{T}^{-1} \otimes \boldsymbol{I}\right)(\boldsymbol{L} \otimes \boldsymbol{Q})\left(\boldsymbol{T}\boldsymbol{T}^{-1} \otimes \boldsymbol{I}\right)\boldsymbol{x}(t) \\
&= \underbrace{\boldsymbol{x}^T(t)\,(\boldsymbol{T} \otimes \boldsymbol{I})}_{\tilde{\boldsymbol{x}}^T(t)}\left(\boldsymbol{T}^{-1} \otimes \boldsymbol{I}\right)(\boldsymbol{L} \otimes \boldsymbol{Q})\,(\boldsymbol{T} \otimes \boldsymbol{I})\underbrace{\left(\boldsymbol{T}^{-1} \otimes \boldsymbol{I}\right)\boldsymbol{x}(t)}_{\tilde{\boldsymbol{x}}(t)} \\
&= \tilde{\boldsymbol{x}}^T(t)\left(\boldsymbol{T}^{-1}\boldsymbol{L}\boldsymbol{T} \otimes \boldsymbol{Q}\right)\tilde{\boldsymbol{x}}(t) \\
&= \tilde{\boldsymbol{x}}^T(t)\,(\operatorname{diag}(\lambda_i) \otimes \boldsymbol{Q})\,\tilde{\boldsymbol{x}}(t) \\
&= \tilde{\boldsymbol{x}}^T(t)\operatorname{diag}(\lambda_i\,\boldsymbol{Q})\,\tilde{\boldsymbol{x}}(t) \\
&= \sum_{i=1}^{N} \lambda_i\,\tilde{\boldsymbol{x}}_i^T(t)\,\boldsymbol{Q}\,\tilde{\boldsymbol{x}}_i(t),
\end{aligned} \tag{3.21}$$

Note that for undirected communication networks, the Laplacian matrix is symmetric and hence has real eigenvalues and orthogonal eigenvectors $\left(\boldsymbol{T}^T = \boldsymbol{T}^{-1}\right)$.

The input transformation (3.20) is applied to the second weighting term in the objective function in (3.15) and yields

$$\begin{aligned}
\boldsymbol{u}^T(t)\,(\boldsymbol{I}\otimes\boldsymbol{R})\,\boldsymbol{u}(t) &= \boldsymbol{u}^T(t)\left(\boldsymbol{T}\boldsymbol{T}^T\otimes\boldsymbol{R}\right)\boldsymbol{u}(t)\\
&= \underbrace{\boldsymbol{u}^T(t)\,(\boldsymbol{T}\otimes\boldsymbol{I})}_{\tilde{\boldsymbol{u}}^T(t)}\,(\boldsymbol{I}\otimes\boldsymbol{R})\,\underbrace{\left(\boldsymbol{T}^{-1}\otimes\boldsymbol{I}\right)\boldsymbol{u}(t)}_{\tilde{\boldsymbol{u}}(t)}\\
&= \sum_{i=1}^{N}\tilde{\boldsymbol{u}}_i^T(t)\,\boldsymbol{R}\,\tilde{\boldsymbol{u}}_i(t).
\end{aligned} \tag{3.22}$$

From the transformations (3.21) and (3.22) it is easy to see that the objective function in (3.15) is given by the following sum

$$\begin{aligned}
J\left(\boldsymbol{u},\,\boldsymbol{x}(0)\right) &= \int_0^\infty \sum_{i=1}^{N}\lambda_i\,\tilde{\boldsymbol{x}}_i^T(t)\,\boldsymbol{Q}\,\tilde{\boldsymbol{x}}_i(t) + \sum_{i=1}^{N}\tilde{\boldsymbol{u}}_i^T(t)\,\boldsymbol{R}\,\tilde{\boldsymbol{u}}_i(t)\,\mathrm{d}t\\
&= \sum_{i=1}^{N}\underbrace{\int_0^\infty \lambda_i\,\tilde{\boldsymbol{x}}_i^T(t)\,\boldsymbol{Q}\,\tilde{\boldsymbol{x}}_i(t) + \tilde{\boldsymbol{u}}_i^T(t)\,\boldsymbol{R}\,\tilde{\boldsymbol{u}}_i(t)\,\mathrm{d}t\,.}_{J_i\left(\tilde{\boldsymbol{u}}_i,\,\tilde{\boldsymbol{x}}_i(0)\right)}
\end{aligned}$$

As the transformed overall system (3.16)

$$\begin{aligned}
\dot{\tilde{\boldsymbol{x}}}(t) &= \left(\boldsymbol{T}^{-1}\otimes\boldsymbol{I}\right)(\boldsymbol{I}\otimes\boldsymbol{A})\,\boldsymbol{x}(t) + \left(\boldsymbol{T}^{-1}\otimes\boldsymbol{I}\right)(\boldsymbol{I}\otimes\boldsymbol{B})\,\boldsymbol{u}(t)\\
&= (\boldsymbol{I}\otimes\boldsymbol{A})\left(\boldsymbol{T}^{-1}\otimes\boldsymbol{I}\right)\boldsymbol{x}(t) + (\boldsymbol{I}\otimes\boldsymbol{B})\left(\boldsymbol{T}^{-1}\otimes\boldsymbol{I}\right)\boldsymbol{u}(t)\\
&= (\boldsymbol{I}\otimes\boldsymbol{A})\,\tilde{\boldsymbol{x}}(t) + (\boldsymbol{I}\otimes\boldsymbol{B})\,\tilde{\boldsymbol{u}}(t)
\end{aligned}$$

is given by decoupled subsystems, the optimization problem (3.43) can be independently solved by N separate optimization problems

$$\begin{aligned}
&\min_{\tilde{\boldsymbol{u}}_i}\; J_i(\tilde{\boldsymbol{u}}_i,\,\tilde{\boldsymbol{x}}_i(0))\\
&s.t.\quad \dot{\tilde{\boldsymbol{x}}}_i(t) = \boldsymbol{A}\,\tilde{\boldsymbol{x}}_i(t) + \boldsymbol{B}\,\tilde{\boldsymbol{u}}_i(t), \qquad \tilde{\boldsymbol{x}}_i(0) = \tilde{\boldsymbol{x}}_{i0}.
\end{aligned}$$

of order n for $i = 1,\,2,\,\ldots,\,N$. □

The decomposition in Lemma 3.1 reveals a significant similarity to the synchronization condition presented in Theorem 3.1. In both cases $N-1$ decoupled subsystems which belong to the eigenvalues λ_i, $(i = 2,\,3,\,\ldots,\,N)$ of the Laplacian matrix $\boldsymbol{L}$ have to be stable for synchronization. This fact can be seen by considering the solution of the decomposed optimization problem (3.17) – (3.18) for $\lambda_1 = 0$, which is trivially given by $\tilde{\boldsymbol{u}}_1(t) = \boldsymbol{0}$. Hence, the dynamics corresponding to the unobservable behavior coincide with the first decoupled subsystem that constitutes the synchronous trajectory $\boldsymbol{y}_{\mathrm{s}}(t)$:

$$\Sigma_{\mathrm{s}} : \begin{cases} \dot{\tilde{\boldsymbol{x}}}_1(t) = \boldsymbol{A}\,\tilde{\boldsymbol{x}}_1(t), & \tilde{\boldsymbol{x}}_1(0) = \dfrac{1}{N}\displaystyle\sum_{i=1}^{N} \boldsymbol{x}_{0i} \\ \boldsymbol{y}_{\mathrm{s}}(t) = \boldsymbol{C}\,\tilde{\boldsymbol{x}}_1(t). \end{cases}$$

3.7.2 Optimal synchronization of completely coupled agents

The solutions of the optimization problems (3.17) – (3.18) for $(i = 2, 3, \ldots, N)$ can be obtained as described in Section 2.3, since $\boldsymbol{Q}$ is symmetric positive definite, $\lambda_i > 0$ and the pair $(\boldsymbol{A}, \boldsymbol{B})$ completely controllable.

Lemma 3.2 (Solution of (3.17) – (3.18))**.** *Let the matrices $\boldsymbol{Q}$ and $\boldsymbol{R}$ be symmetric positive definite, $\lambda_i > 0$, $(i = 2, 3, \ldots, N)$ and the pair $(\boldsymbol{A}, \boldsymbol{B})$ completely controllable. Then the solutions of the decomposed optimization problems* (3.17) – (3.18) *are given by*

$$\tilde{\boldsymbol{u}}_i(t) = -\boldsymbol{R}^{-1}\boldsymbol{B}^T\boldsymbol{P}_i\,\tilde{\boldsymbol{x}}_i(t), \qquad i = 1, 2, \ldots, N \tag{3.23}$$

with $\boldsymbol{P}_1 = \boldsymbol{O}$. The matrices $\boldsymbol{P}_i$, $(i = 2, 3, \ldots, N)$ are the symmetric positive definite solutions of the following $N-1$ algebraic riccati equations

$$\boldsymbol{A}^T\boldsymbol{P}_i + \boldsymbol{P}_i\boldsymbol{A} - \boldsymbol{P}_i\boldsymbol{B}\boldsymbol{R}^{-1}\boldsymbol{B}^T\boldsymbol{P}_i + \lambda_i\boldsymbol{Q} = \boldsymbol{O}. \tag{3.24}$$

Proof. The proof follows directly from the LQR methods described in Section 2.3. □

Finally, the solution of the optimization problem (3.17) – (3.18) is obtained by the combination of the results of Lemma 3.1 and Lemma 3.2.

Theorem 3.2 (Optimal synchronizing state feedback)**.** *The solution of the optimization problem* (3.17) – (3.18) *is given by*

$$\boldsymbol{u}(t) = -\left(\boldsymbol{T} \otimes \boldsymbol{I}\right)\mathrm{diag}\left(\boldsymbol{R}^{-1}\boldsymbol{B}^T\boldsymbol{P}_i\right)\left(\boldsymbol{T}^{-1} \otimes \boldsymbol{I}\right)\boldsymbol{x}(t), \tag{3.25}$$

where $\boldsymbol{T}$ is the eigenvector matrix of the Laplacian matrix $\boldsymbol{L}$. The matrix $\boldsymbol{P}_1 = \boldsymbol{O}$ and the remaining matrices $\boldsymbol{P}_i$, $(i = 2, 3, \ldots, N)$ are the symmetric positive definite solutions of the algebraic riccati equations defined in (3.24).

Proof. In terms of the representation of the overall system, the solution of the decomposed optimization problems (3.17) – (3.17) is summarized by

$$\tilde{\boldsymbol{u}}(t) = -\operatorname{diag}\left(\boldsymbol{R}^{-1}\boldsymbol{B}^T\boldsymbol{P}_i\right)\tilde{\boldsymbol{x}}(t), \tag{3.26}$$

where the matrices $\boldsymbol{P}_i$, $(i = 2, 3, \ldots, N)$ are the symmetric, positive definite solutions of (3.24). In reference to $\tilde{\boldsymbol{u}}_1 = \boldsymbol{0}$, the matrix $\boldsymbol{P}_1$ is defined to be zero. The proof of the theorem is completed by transforming the solution (3.26) of the decomposed optimization problem with (3.19) and (3.20) back to the original representation:

$$\begin{aligned}
(\boldsymbol{T}\otimes\boldsymbol{I})^{-1}\boldsymbol{u}(t) &= -\operatorname{diag}\left(\boldsymbol{R}^{-1}\boldsymbol{B}^T\boldsymbol{P}_i\right)\left(\boldsymbol{T}^{-1}\otimes\boldsymbol{I}\right)\boldsymbol{x}(t)\\
\boldsymbol{u}(t) &= -\left(\boldsymbol{T}\otimes\boldsymbol{I}\right)\operatorname{diag}\left(\boldsymbol{R}^{-1}\boldsymbol{B}^T\boldsymbol{P}_i\right)\left(\boldsymbol{T}^{-1}\otimes\boldsymbol{I}\right)\boldsymbol{x}(t).
\end{aligned}$$

□

Whereas the optimization problem (3.17) – (3.18) has the order Nn, the decomposition obtained in Lemma 3.1 leads to a reduction of the computational load, since now, the solution of $N-1$ algebraic riccati equations of the order n of a single agent has to be obtained. Note that depending on the multiplicity of the eigenvalues λ_i, $(i = 1, 2, \ldots, N)$ the complexity of designing the optimal feedback can be reduced further.

According to the structure of the networked controller, Theorem 3.2 shows that the solution of the optimization problem (3.17) – (3.18) is given by the state feedback (3.25), which is different from the desired form of the networked controller presented in (3.8). The optimal synchronizing feedback (3.25) exhibits the desired structure in (3.8) if and only if the matrices $\boldsymbol{P}_i$ are linearly dependent with respect to the eigenvalues λ_i:

$$\boldsymbol{P}_i = \lambda_i\,\boldsymbol{P}, \qquad i = 1, 2, \ldots, N. \tag{3.27}$$

In that case, the feedback (3.25) can be rewritten as

$$\begin{aligned}
\boldsymbol{u}(t) &= -\left(\boldsymbol{T}\otimes\boldsymbol{I}\right)\operatorname{diag}\left(\lambda_i\,\boldsymbol{R}^{-1}\boldsymbol{B}^T\boldsymbol{P}\right)\left(\boldsymbol{T}^{-1}\otimes\boldsymbol{I}\right)\boldsymbol{x}(t)\\
&= -\left(\boldsymbol{T}\otimes\boldsymbol{I}\right)\left(\operatorname{diag}\left(\lambda_i\right)\otimes\boldsymbol{R}^{-1}\boldsymbol{B}^T\boldsymbol{P}\right)\left(\boldsymbol{T}^{-1}\otimes\boldsymbol{I}\right)\boldsymbol{x}(t)\\
&= -\left(\boldsymbol{T}\operatorname{diag}\left(\lambda_i\right)\boldsymbol{T}^{-1}\otimes\boldsymbol{R}^{-1}\boldsymbol{B}^T\boldsymbol{P}\right)\boldsymbol{x}(t)\\
&= -\left(\boldsymbol{L}\otimes\boldsymbol{K}\right)\boldsymbol{x}(t)
\end{aligned}$$

with $\boldsymbol{K} = \boldsymbol{R}^{-1}\boldsymbol{B}^T\boldsymbol{P}$. Since the algebraic riccati equations (3.24) are quadratic, it is obvious that (3.27) is in general not fulfilled. Hence, the design method described in this section typically results in a dense controller matrix (3.25). Consequently, solving the optimization problem (3.17) – (3.18) for a given Laplacian matrices $\boldsymbol{L}$ that represents a non-complete network structure produces a controller (3.25) which requires an all-to-all coupling between the agents. Hence, the resulting optimal synchronizing controller does not fulfill the requirement on the structure of the networked controller as shown in Fig. 3.1, namely, the separation into identical local controllers and the communication network described by a Laplacian matrix.

Assuming that the networked controller (3.8) is composed of a communication network which couples all agents with all others, then the Laplacian matrix is described by

$$\boldsymbol{L}_{\mathcal{K}} = \begin{pmatrix} N-1 & -1 & \cdots & -1 \\ -1 & N-1 & \ddots & \vdots \\ \vdots & \ddots & \ddots & -1 \\ -1 & \cdots & -1 & N-1 \end{pmatrix}.$$

Note that the Laplacian matrix $\boldsymbol{L}_{\mathcal{K}}$ has two distinct eigenvalues $\lambda_1(\boldsymbol{L}_{\mathcal{K}}) = 0$ and the eigenvalue $\lambda_2(\boldsymbol{L}_{\mathcal{K}}) = N$ of multiplicity $N-1$. The decomposition in Lemma 3.1 and the corresponding solution in Lemma 3.2 show that in this case the optimization problem (3.17) – (3.18) is solved by

$$\begin{aligned} \tilde{\boldsymbol{u}}_i(t) &= -\boldsymbol{R}^{-1}\boldsymbol{B}^T\boldsymbol{P}_2\,\tilde{\boldsymbol{x}}_i(t), \qquad i = 2, 3, \ldots, N \\ &= -N\,\boldsymbol{R}^{-1}\boldsymbol{B}^T\frac{1}{N}\,\boldsymbol{P}_2\,\tilde{\boldsymbol{x}}_i(t), \end{aligned} \tag{3.28}$$

where $\boldsymbol{P}_1 = \boldsymbol{O}$ and $\boldsymbol{P}_2$ is the symmetric positive definite solution of the Riccati equation

$$\boldsymbol{A}^T\boldsymbol{P}_2 + \boldsymbol{P}_2\boldsymbol{A} - \boldsymbol{P}_2\boldsymbol{B}\boldsymbol{R}^{-1}\boldsymbol{B}^T\boldsymbol{P}_2 + N\boldsymbol{Q} = \boldsymbol{O}.$$

Hence, the complexity of determining the optimal feedback reduces from order Nn to the order n of a single agent. According to (3.28) the overall representation, the solution of the decomposed optimization problem can be rewritten as

$$\begin{aligned} \tilde{\boldsymbol{u}}(t) &= -\operatorname{diag}\left(\lambda_i(\boldsymbol{L}_{\mathcal{K}})\,\boldsymbol{R}^{-1}\boldsymbol{B}^T\frac{1}{N}\boldsymbol{P}_2\right)\tilde{\boldsymbol{x}}(t) \\ &= -\left(\operatorname{diag}(\lambda_i(\boldsymbol{L}_{\mathcal{K}})) \otimes \boldsymbol{R}^{-1}\boldsymbol{B}^T\frac{1}{N}\boldsymbol{P}_2\right)\tilde{\boldsymbol{x}}(t). \end{aligned}$$

Finally, the optimal synchronizing feedback that is obtained by application of the transformations (3.19)– (3.20) is given by

$$\boldsymbol{u}(t) = -\left(\boldsymbol{T} \otimes \boldsymbol{I}\right)\left(\operatorname{diag}\left(\lambda_i\left(\boldsymbol{L}_{\mathcal{K}}\right)\right) \otimes \boldsymbol{R}^{-1}\boldsymbol{B}^T\frac{1}{N}\boldsymbol{P}_2\right)\left(\boldsymbol{T}^{-1} \otimes \boldsymbol{I}\right)\boldsymbol{x}(t)$$

$$\boldsymbol{u}(t) = -\left(\boldsymbol{T}\operatorname{diag}\left(\lambda_i\left(\boldsymbol{L}_{\mathcal{K}}\right)\right)\boldsymbol{T}^{-1} \otimes \boldsymbol{R}^{-1}\boldsymbol{B}^T\frac{1}{N}\boldsymbol{P}_2\right)\boldsymbol{x}(t)$$

$$\boldsymbol{u}(t) = -\Big(\boldsymbol{L}_{\mathcal{K}} \otimes \underbrace{\boldsymbol{R}^{-1}\boldsymbol{B}^T\frac{1}{N}\boldsymbol{P}_2}_{\boldsymbol{K}}\Big)\boldsymbol{x}(t). \tag{3.29}$$

As shown in the previous calculations, the optimal synchronizing feedback reveals the desired structure given in (3.8) only in the special case of a complete agent coupling.

3.8 Approximate method for the design of synchronizing controllers for non-complete networks

To overcome the limitation of a complete agent coupling, an approximation method is developed that ensures synchronization of the agents for any connected communication network. The obtained approximate optimal feedback does clearly not minimize the objective function in eqn. (3.15), but satisfies the requirement (3.1) on asymptotic synchronization. For the special case of a complete communication network it is shown that the designed approximate networked controller is also optimal.

Comparing the optimal control law (3.25) and the desired control law (3.8) shows with

$$\begin{aligned}\boldsymbol{u}(t) &= -\left(\boldsymbol{L} \otimes \boldsymbol{K}\right)\boldsymbol{x}(t)\\ &= -\left(\boldsymbol{T}\boldsymbol{T}^{-1} \otimes \boldsymbol{I}\right)\left(\boldsymbol{L} \otimes \boldsymbol{K}\right)\left(\boldsymbol{T}\boldsymbol{T}^{-1} \otimes \boldsymbol{I}\right)\boldsymbol{x}(t)\\ &= -\left(\boldsymbol{T} \otimes \boldsymbol{I}\right)\left(\operatorname{diag}\left(\lambda_i\right) \otimes \boldsymbol{K}\right)\left(\boldsymbol{T}^{-1} \otimes \boldsymbol{I}\right)\boldsymbol{x}(t)\end{aligned}$$

that the desired networked controller (3.8) is optimal if and only if

$$\lambda_i\boldsymbol{K} = \boldsymbol{R}^{-1}\boldsymbol{B}^T\boldsymbol{P}_i, \qquad j = 1, 2, \ldots, N \tag{3.30}$$

holds. As already described in Section 3.7.2, it can be seen from the substitution $\boldsymbol{K} = \boldsymbol{R}^{-1}\boldsymbol{B}^T\boldsymbol{P}$ with the invertible matrix $\boldsymbol{P} \in \mathbb{R}^{n\times n}$ that the requirement (3.30) cannot be generally satisfied, since

$$\lambda_i \boldsymbol{P} \neq \boldsymbol{P}_i, \qquad i = 2, 3, \ldots, N$$

or equally

$$\boldsymbol{X}_i \neq \frac{1}{\lambda_i}\boldsymbol{X} \tag{3.31}$$

with $\boldsymbol{X}_i = \boldsymbol{P}_i^{-1}$ and $\boldsymbol{X} = \boldsymbol{P}^{-1}$. In view of relation (3.31) a performance index is derived which is suitable for the design of the matrix $\boldsymbol{P}$. Namely, the matrix $\boldsymbol{X} = \boldsymbol{P}^{-1}$ is determined by the solution of the minimization problem

$$\min_{\boldsymbol{X}} S = \sum_{i=2}^{N} \left\| \boldsymbol{X}_i - \frac{1}{\lambda_i}\boldsymbol{X} \right\|_{\mathrm{F}} \tag{3.32}$$

with S representing the sum of squared residuals (see [125] and the references therein for an overview on generalized least-squares methods).

The following theorem shows under which condition the approximate synchronizing networked controller exists.

Theorem 3.3 (Approximate synchronizing state feedback)**.** *If the second largest Laplacian eigenvalue λ_2 satisfies the bound*

$$\lambda_2 \geq \frac{1}{2}\left(\sum_{k=2}^{N} 1/\lambda_k^2\right)^{-1} \sum_{i=2}^{N} \frac{1}{\lambda_i} \tag{3.33}$$

then the networked controller

$$\boldsymbol{u}(t) = -\left(\boldsymbol{L} \otimes \boldsymbol{R}^{-1}\boldsymbol{B}^T\boldsymbol{X}^{-1}\right)\boldsymbol{x}(t) \tag{3.34}$$

with

$$\boldsymbol{X} = \sum_{i=2}^{N} \underbrace{\left(\sum_{k=2}^{N} 1/\lambda_k^2\right)^{-1} \frac{1}{\lambda_i}}_{\alpha_i} \boldsymbol{X}_i \tag{3.35}$$

asymptotically synchronizes the networked multi-agent system. The matrix (3.35) *is the solution of the minimization problem* (3.32).

Proof. The least-squares method described in [125] allows to directly determine the solution of the minimization problem (3.32), which is given by (3.35). The following computations show that (3.35) solves the synchronization problem if and only if the second eigenvalue of the Laplacian matrix λ_2 is large enough.

Since all $\alpha_i > 0$, $(i = 2, 3, \ldots, N)$ and $\boldsymbol{X}_i = \boldsymbol{X}_i^T \succ 0$, $\boldsymbol{X}$ in (3.35) is symmetric positive definite too. From (3.35) and the sum of the algebraic riccati equations (3.24) multiplied by α_i, $(i = 2, \ldots, N)$ it is easy to see that $\boldsymbol{X}$ can be expressed as the solution of the Lyapunov equation

$$\boldsymbol{X}\boldsymbol{A}^T + \boldsymbol{A}\boldsymbol{X} - \sum_{i=2}^{N} \alpha_i \boldsymbol{B}\boldsymbol{R}^{-1}\boldsymbol{B}^T + \underbrace{\sum_{i=2}^{N} \alpha_i \boldsymbol{X}_i \boldsymbol{Q} \boldsymbol{X}_i}_{\tilde{\boldsymbol{Q}}} = \boldsymbol{O}. \tag{3.36}$$

It is obvious that eqn. (3.36) can be represented in terms of the synchronization condition given in Theorem 3.1:

$$\begin{aligned} \boldsymbol{X}\left(\boldsymbol{A} - \lambda_i \boldsymbol{B}\boldsymbol{R}^{-1}\boldsymbol{B}^T\boldsymbol{X}^{-1}\right)^T + \left(\boldsymbol{A} - \lambda_i \boldsymbol{B}\boldsymbol{R}^{-1}\boldsymbol{B}^T\boldsymbol{X}^{-1}\right)\boldsymbol{X} + \\ \left(2\lambda_i - \sum_{j=2}^{N} \alpha_j\right)\boldsymbol{B}\boldsymbol{R}^{-1}\boldsymbol{B}^T + \tilde{\boldsymbol{Q}} = \boldsymbol{O}. \end{aligned} \tag{3.37}$$

Equation (3.37) shows that the synchronization condition (3.11) is fulfilled if

$$\left(2\lambda_i - \sum_{j=2}^{N} \alpha_j\right)\boldsymbol{B}\boldsymbol{R}^{-1}\boldsymbol{B}^T + \tilde{\boldsymbol{Q}} \succ 0, \qquad i = 2, 3, \ldots, N. \tag{3.38}$$

Since $\boldsymbol{Q}$ is symmetric positive definite and the matrices $\boldsymbol{X}_j$, $(j = 2, \ldots, N)$ are symmetric, it follows that $\tilde{\boldsymbol{Q}} = \tilde{\boldsymbol{Q}}^T \succ 0$. Hence, eqn. (3.38) is true if

$$2\lambda_i - \sum_{j=2}^{N} \alpha_j \geq 0. \tag{3.39}$$

Clearly, since λ_2 is the second smallest eigenvalue it represents the worst case for eqn. (3.39) to be positive definite. □

Theorem 3.3 shows that the solution of the optimization problem (3.15) – (3.16) can be used for the design of networked controllers (3.8) for arbitrary communication networks $\boldsymbol{L}$ with λ_2 satisfying the bound (3.33). The optimization problem (3.32) is chosen to minimize the deviation between the optimal networked controller provided by Theorem 3.2 and the desired structure in eqn. (3.8). Since the networked controller is obtained by approximating the optimal solution it does not minimize the objective function in (3.15) but, nevertheless, satisfies the requirement on asymptotic synchronization if (3.33) holds.

In the case where (3.33) is not satisfied by the communication network $\boldsymbol{L}$, the tuning factor

ϵ can be used to ensure asymptotic synchronization by increasing the open-loop gain:

$$\boldsymbol{u}(t) = -\epsilon\,(\boldsymbol{L} \otimes \underbrace{\boldsymbol{R}^{-1}\boldsymbol{B}^T\boldsymbol{X}^{-1}}_{\boldsymbol{K}})\boldsymbol{x}(t). \tag{3.40}$$

Corollary 3.2. *The networked controller* (3.40) *with* $\boldsymbol{X}$ *given by* (3.35) *and*

$$\epsilon \geq \frac{1}{2\,\lambda_2}\sum_{j=2}^{N}\alpha_j$$

asymptotically synchronizes the networked multi-agent system (3.9)

Proof. Consider eqn. (3.37) with respect to the modified feedback (3.40):

$$\boldsymbol{X}\,(\boldsymbol{A}-\epsilon\,\lambda_i\,\boldsymbol{BK})^T + (\boldsymbol{A}-\epsilon\,\lambda_i\,\boldsymbol{BK})\,\boldsymbol{X} + \left(2\epsilon\,\lambda_i - \sum_{j=2}^{N}\alpha_j\right)\boldsymbol{BR}^{-1}\boldsymbol{B}^T + \tilde{\boldsymbol{Q}} = \boldsymbol{O}. \tag{3.41}$$

Equation (3.41) shows that the matrices $\boldsymbol{A}-\epsilon\,\lambda_i\,\boldsymbol{BK}$, $(i=2,\,3,\,\ldots,\,N)$ are Hurwitz and the networked multi-agent system correspondingly synchronized if

$$\left(2\epsilon\,\lambda_i - \sum_{j=2}^{N}\alpha_j\right)\boldsymbol{BR}^{-1}\boldsymbol{B}^T + \tilde{\boldsymbol{Q}} \succ 0. \tag{3.42}$$

Equation (3.42) is fulfilled for the bound given in (3.2). □

The following corollary shows that the approximation is tight in the sense that the approximate solution is, in the special case of a complete graph, also optimal.

Corollary 3.3. *Let the communication network of the multi-agent system* (3.9) *be described by the Laplacian matrix* $\boldsymbol{L}_{\mathcal{K}}$*. Then the networked controller provided by Theorem 3.3 is equal to the optimal solution given in Theorem 3.2.*

Proof. With respect to the eigenvalues of the Laplacian $\boldsymbol{L}_{\mathcal{K}}$, $\lambda_1 = 0$, $\lambda_2 = \ldots = \lambda_N = N$ it is easy to see that the solutions of (3.24) are identical ($\boldsymbol{P}_i = \boldsymbol{P}$) and eqn. (3.30) fulfilled with $\boldsymbol{K} = \boldsymbol{R}^{-1}\boldsymbol{B}^T\,1/N\,\boldsymbol{P}$. Hence, the corollary is proved by virtue of Theorem 3.3. □

3.9 Optimal synchronization in non-complete networks

In the previous two sections the solution of the optimization problem (3.15) – (3.16) is presented and it is shown that an all-to-all coupling of the agents is required for optimal synchronization. Based on the optimal synchronizing controller (3.25) an approximate method is derived which is suitable for the design of networked controllers for non-complete communication networks. Since the approximate solution does not minimize the objective function, the optimization problem (3.15) – (3.16) is modified in this section to consider the structure of the communication network as an additional constraint. Thus, the solution of the modified optimization problem fulfills the requirement on optimal synchronization.

The modification of the optimization problem (3.15) – (3.16) is achieved by first dropping the dependence of the objective function on the initial condition $\boldsymbol{x}_0$ and, secondly, by considering the given communication network as a constraint. Now, the solution of the resulting optimization problem is obtained and it is shown that it is computationally expensive to solve. For this reason an algorithm is proposed that uses a gradient descent method for the design of a controller matrix $\boldsymbol{K}$ which minimizes the modified objective function for a given initial synchronizing controller and hence fulfills the requirement on optimal synchronization.

3.9.1 LQ-like design problem for synchronization

The design of the networked controller (3.8) is solved by a parametric optimization problem, where the controller matrix $\boldsymbol{K}$ should minimize the performance index in (3.15). The design of constrained LQ-based controllers was first investigated by Levine and Athans in [126] and has been intensively studied in the last decades.

The optimization problem concerned for the design of optimal synchronizing controllers in non-complete networks is given by

$$\min_{\boldsymbol{K}} \ \mathrm{E}\left(J(\boldsymbol{u}(t),\, \boldsymbol{x}_0)\right) \tag{3.43}$$

$$s.t. \ \dot{\boldsymbol{x}}(t) = (\boldsymbol{I} \otimes \boldsymbol{A})\, \boldsymbol{x}(t) + (\boldsymbol{I} \otimes \boldsymbol{B})\, \boldsymbol{u}(t),\ \boldsymbol{x}(0) = \boldsymbol{x}_0, \tag{3.44}$$

$$\boldsymbol{u}(t) = -\left(\boldsymbol{L} \otimes \boldsymbol{K}\right) \boldsymbol{x}(t) \tag{3.45}$$

with $\mathrm{E}\,(.)$ representing the expected value of the objective function

$$J(\boldsymbol{u}(t),\,\boldsymbol{x}_0) = \int_0^\infty \boldsymbol{x}^T(t)\,(\boldsymbol{L}\otimes\boldsymbol{Q})\,\boldsymbol{x}(t) + \boldsymbol{u}^T(t)\boldsymbol{u}(t)\,\mathrm{d}t. \tag{3.46}$$

The weighting matrix $\boldsymbol{R}$ is chosen to be $\boldsymbol{R}=\boldsymbol{I}$ for simplicity.

Substituting the constraints (3.45) and (3.10) into the objective function (3.46) gives

$$J(\boldsymbol{K},\boldsymbol{x}_0) = \int_0^\infty \boldsymbol{x}_0^T\,\boldsymbol{\Phi}^T(t)\Big(\,(\boldsymbol{L}\otimes\boldsymbol{Q}) + (\boldsymbol{L}\otimes\boldsymbol{K})^T\,(\boldsymbol{L}\otimes\boldsymbol{K})\,\Big)\boldsymbol{\Phi}(t)\,\boldsymbol{x}_0\,\mathrm{d}t \tag{3.47}$$

$$= \boldsymbol{x}_0^T \underbrace{\int_0^\infty \boldsymbol{\Phi}^T(t)\Big(\,(\boldsymbol{L}\otimes\boldsymbol{Q}) + \big(\boldsymbol{L}^2\otimes\boldsymbol{K}^T\boldsymbol{K}\big)\,\Big)\boldsymbol{\Phi}(t)\,\mathrm{d}t}_{\boldsymbol{F}\,(\boldsymbol{K})}\;\boldsymbol{x}_0 \tag{3.48}$$

with

$$\boldsymbol{\Phi}(t) = \mathrm{e}^{(\boldsymbol{I}\otimes\boldsymbol{A}\,-\,\boldsymbol{L}\otimes\boldsymbol{B}\boldsymbol{K})\,t}.$$

Equation (3.47) reveals the dependence of the performance index $J(\boldsymbol{K},\boldsymbol{x}_0)$ on both, the control matrix $\boldsymbol{K}$ and the initial condition $\boldsymbol{x}_0$. Note that the solution of the optimization problem $\min_{\boldsymbol{K}} J(\boldsymbol{K},\boldsymbol{x}_0)$ would depend on the initial condition, which is not desired when designing local controllers $\boldsymbol{K}$. A well established method to remove this dependency is to assume that $\boldsymbol{x}_0$ is uniformly distributed over a unit sphere and then to compute the expected value of the objective function $J(\boldsymbol{K},\boldsymbol{x}_0)$, as considered in eqn. (3.43).

The expected value of the objective function (3.46) can be expressed in terms of the trace, since

$$\begin{aligned}
\mathrm{E}\,(J\,(\boldsymbol{K},\boldsymbol{x}_0)) &= \mathrm{E}\Big(\,\mathrm{tr}\,\big(\boldsymbol{x}_0^T\,\boldsymbol{F}\,(\boldsymbol{K})\,\boldsymbol{x}_0\big)\Big)\\
&= \mathrm{E}\Big(\,\mathrm{tr}\,\big(\boldsymbol{F}\,(\boldsymbol{K})\,\boldsymbol{x}_0\,\boldsymbol{x}_0^T\big)\Big)\\
&= \mathrm{tr}\,\Big(\boldsymbol{F}\,(\boldsymbol{K})\,\mathrm{E}\,\big(\boldsymbol{x}_0\,\boldsymbol{x}_0^T\big)\Big)\\
&= \sigma\,\mathrm{tr}\,\Big(\boldsymbol{F}\,(\boldsymbol{K})\Big)
\end{aligned} \tag{3.49}$$

with $\mathrm{E}\,\big(\boldsymbol{x}_0\,\boldsymbol{x}_0^T\big) = \sigma\boldsymbol{I}$ and $\sigma = \mathrm{E}\,(\boldsymbol{x}_{0i}^2)$. Equation (3.49) shows that minimizing the trace of the matrix $\boldsymbol{F}\,(\boldsymbol{K})$ is similar to minimizing the expected value of the objective function (3.47).

This approach is a standard method in linear and quadratic optimization [127].

Hence, the optimization problem (3.43) – (3.45) can be rewritten as

$$\min_{\boldsymbol{K}} \tilde{J}(\boldsymbol{K}) \tag{3.50}$$

with $\tilde{J}(\boldsymbol{K}) = \operatorname{tr}\Big(\boldsymbol{F}(\boldsymbol{K})\Big)$.

3.9.2 Solution of the optimization problem

The solution of the optimization problem (3.50) is obtained by application of the calculus of variations. Since the objective function (3.49) is continuously differentiable, the first-order necessary condition for optimality is

$$\nabla \tilde{J}(\boldsymbol{K})\Big|_{\boldsymbol{K}^*} = \boldsymbol{O}. \tag{3.51}$$

The following lemma shows that the gradient of the objective function $\tilde{J}(\boldsymbol{K})$ involves the solution of $2(N-1)$ Lyapunov equations of the order n.

Lemma 3.3. *The gradient matrix $\nabla \tilde{J}(\boldsymbol{K})$ of the objective function $\tilde{J}(\boldsymbol{K}) = \operatorname{tr}\Big(\boldsymbol{F}(\boldsymbol{K})\Big)$ with $\boldsymbol{F}(\boldsymbol{K})$ given in* (3.48) *is obtained by*

$$\nabla \tilde{J}(\boldsymbol{K}) = 2 \sum_{i=2}^{N} \lambda_i \left(\lambda_i \boldsymbol{K} - \boldsymbol{B}^T \boldsymbol{P}_i\right) \boldsymbol{X}_i, \tag{3.52}$$

where $\boldsymbol{P}_i$ and $\boldsymbol{X}_i$, $(i = 2, \ldots, N)$ are symmetric positive definite solutions of the Lyapunov equations

$$(\boldsymbol{A} - \lambda_i \boldsymbol{B}\boldsymbol{K})^T \boldsymbol{P}_i + \boldsymbol{P}_i (\boldsymbol{A} - \lambda_i \boldsymbol{B}\boldsymbol{K}) + \lambda_i^2 \boldsymbol{K}^T\boldsymbol{K} + \lambda_i \boldsymbol{Q} = \boldsymbol{O} \tag{3.53}$$

$$(\boldsymbol{A} - \lambda_i \boldsymbol{B}\boldsymbol{K}) \boldsymbol{X}_i + \boldsymbol{X}_i (\boldsymbol{A} - \lambda_i \boldsymbol{B}\boldsymbol{K})^T + \boldsymbol{I} = \boldsymbol{O}. \tag{3.54}$$

Proof. The proof is given in the appendix and is motivated by the results presented in [126]. □

Based on Lemma 3.3 a necessary and sufficient condition for optimal synchronization, in the sense of minimizing the objective function $J(\boldsymbol{K})$, is derived.

Theorem 3.4 (Necessary condition for optimal synchronization). *The controller matrix $\boldsymbol{K}$ fulfills the requirement on optimal synchronization for a given Laplacian matrix $\boldsymbol{L}$ if and only if $\boldsymbol{K}$ is the solution of*

$$\boldsymbol{K} = \boldsymbol{B}^T \sum_{i=2}^{N} \lambda_i \boldsymbol{P}_i \boldsymbol{X}_i \left(\sum_{j=2}^{N} \lambda_j^2 \boldsymbol{X}_j \right)^{-1} \tag{3.55}$$

with $\boldsymbol{P}_i = \boldsymbol{P}_i^T \succ 0$ and $\boldsymbol{X}_i = \boldsymbol{X}_i^T \succ 0$ obtained from eqn. (3.53) and (3.54).

Proof. The necessary condition in (3.51) and the corresponding gradient matrix in (3.52) show that eqn. (3.55) holds if the term $\bar{\boldsymbol{X}} = \sum_{j=2}^{N} \lambda_j^2 \boldsymbol{X}_j$ is invertible. Since the matrices $\boldsymbol{X}_i$ are positive definite and $\lambda_i > 0$ for all $i = 2, \ldots, N$, $\bar{\boldsymbol{X}}$ is positive definite and consequently invertible, too. □

The Lyapunov equations (3.53) and (3.54) reveal that the matrices $\tilde{\boldsymbol{A}}_i = \boldsymbol{A} - \lambda_i \boldsymbol{B} \boldsymbol{K}$, $(i = 2, 3, \ldots, N)$ are Hurwitz, since $\boldsymbol{I}$ and $\tilde{\boldsymbol{Q}}_i = \boldsymbol{Q} + \lambda_i \boldsymbol{K}^T \boldsymbol{K}$ are symmetric positive definite. It follows from $\boldsymbol{Q}^T = \boldsymbol{Q} \succ 0$ and $\lambda_i > 0$, $(i = 2, 3, \ldots, N)$ that $\tilde{\boldsymbol{Q}}_i^T = \tilde{\boldsymbol{Q}}_i \succ 0$ and (as required by Theorem 3.4) $\boldsymbol{P}_i^T = \boldsymbol{P}_i \succ 0$. This shows that any networked controller (3.8) satisfying Theorem 3.4 will obviously asymptotically synchronize the networked multi-agent system (3.9).

In view of the results in Section 3.7.2, the solution of the optimization problem (3.50) has to be equal to (3.29) in the case of a complete agent coupling, which is described by the Laplacian matrix $\boldsymbol{L}_{\mathcal{K}}$. Given the eigenvalues of the Laplacian matrix $\boldsymbol{L}_{\mathcal{K}}$, it is evident from the Lyapunov equation (3.53), (3.54) and the controller matrix (3.29) that $\boldsymbol{P}_i = \boldsymbol{P}$, $\boldsymbol{X}_i = \boldsymbol{X}$ for $i = 2, 3, \ldots, N$ and

$$\boldsymbol{K} = \boldsymbol{B}^T \sum_{i=2}^{N} N \boldsymbol{P} \boldsymbol{X} \left(\sum_{j=2}^{N} N^2 \boldsymbol{X} \right)^{-1} = \boldsymbol{B}^T \frac{1}{N} \boldsymbol{P}.$$

holds. This example shows that the objective functions $J(\boldsymbol{u}, \boldsymbol{x}_0)$ and $\tilde{J}(\boldsymbol{K})$ are very similar.

Since the optimization problem (3.50) is non-convex, the global optimal solution is difficult to obtain. The substitution of (3.55) into (3.53) and (3.54) shows that the solution of non-linearly coupled Lyapunov equations has to be obtained for the design of a locally optimal controller matrix $\boldsymbol{K}$. Thus, even the search for candidates is a computationally expensive problem. For this reason, a gradient-based algorithm is proposed for the design of locally optimal networked controllers.

3.9.3 Algorithm for the design of optimal synchronizing controllers

Lemma 3.3 provides equations for the calculation of the gradient of the objective function $\tilde{J}(\boldsymbol{K})$ which are used in this section for the determination of the controller matrix (3.55) to solve the optimization problem (3.50) in the sense of a local minimum. The idea behind the algorithm is based on the use of a gradient descent method to stepwise improve the controller matrix with respect to the value of the objective function $\tilde{J}(\boldsymbol{K})$. Note that for the computation of the gradient (3.52) a controller matrix $\boldsymbol{K}$ is required that has to satisfy asymptotic synchronization of the networked agents. Hence for a given initial controller matrix $\boldsymbol{K}$ which satisfies the requirement on asymptotic synchronization, the algorithm produces a controller matrix $\boldsymbol{K}$ which additionally satisfies the requirement on optimal synchronization.

As shown in Section 3.7, there always exists an initial controller matrix $\boldsymbol{K}$ which asymptotically synchronizes the networked multi-agent system (3.9). Other design methods can be found in [46] or [84].

Algorithm 3.1. *Optimal synchronizing networked controller*

Given: $\boldsymbol{A}$, $\boldsymbol{B}$, $\boldsymbol{L}$, $\gamma \in \mathbb{R}^+$ *and a controller matrix* $\boldsymbol{K}$ *which satisfies the requirement (3.1)*
Initialize: $\boldsymbol{K}_1 = \boldsymbol{K}$, $j = 1$

1. *Compute the gradient* $\nabla \tilde{J}(\boldsymbol{K}_j)$ *by using Lemma 3.3.*

2. *Choose a step-size* $a \in \mathbb{R}^+$ *such that* $\tilde{J}(\boldsymbol{K}_{j+1}) < \tilde{J}(\boldsymbol{K}_j)$ *with* $\boldsymbol{K}_{j+1} = \boldsymbol{K}_j - a\nabla \tilde{J}(\boldsymbol{K}_j)$.

3. *If* $\left\| \nabla \tilde{J}(\boldsymbol{K}_{j+1}) \right\| < \gamma$ *stop. Otherwise set* $j = j + 1$ *and go back to Step 1.*

Result: *Suboptimal controller matrix* $\boldsymbol{K} = \boldsymbol{K}_{j+1}$.

The step-size a can be chosen using any standard step-size selection method. Furthermore, the algorithm has the following property.

Lemma 3.4. *Let* $\boldsymbol{K}$ *be a synchronizing controller matrix satisfying (3.1). Then there always exists a step-size* $a \in \mathbb{R}^+$ *such that* $\tilde{J}(\boldsymbol{K} - a\nabla \tilde{J}(\boldsymbol{K})) \leq \tilde{J}(\boldsymbol{K})$. *The equality* $\tilde{J}(\boldsymbol{K} - a\nabla \tilde{J}(\boldsymbol{K})) = \tilde{J}(\boldsymbol{K})$ *is satisfied if and only if* $\nabla \tilde{J}(\boldsymbol{K}) = 0$.

Proof. The proof of the lemma is included in the Appendix. □

3.10 Application example: Synchronization of a vehicle platoon

As an application example, the synchronization of the lateral dynamics of four mobile robots of the experimental plant SAMS is considered. The model of the mobile robots is given by (2.25). Figure 2.5 illustrates the control aim, which is to synchronize the lateral dynamics of the agents according to the initial conditions

$$\boldsymbol{x}_{01} = \begin{pmatrix} 140 \\ 0 \end{pmatrix}, \quad \boldsymbol{x}_{02} = \begin{pmatrix} 0 \\ 0 \end{pmatrix}, \quad \boldsymbol{x}_{03} = \begin{pmatrix} 0 \\ 0 \end{pmatrix}, \quad \boldsymbol{x}_{04} = \begin{pmatrix} 0 \\ 0 \end{pmatrix}. \tag{3.56}$$

Optimal synchronization with full state feedback

First, the results of Section 3.7 are used for the design of the optimal synchronizing state feedback for the interconnection structures shown in Fig. 3.3. The synchronizing optimal

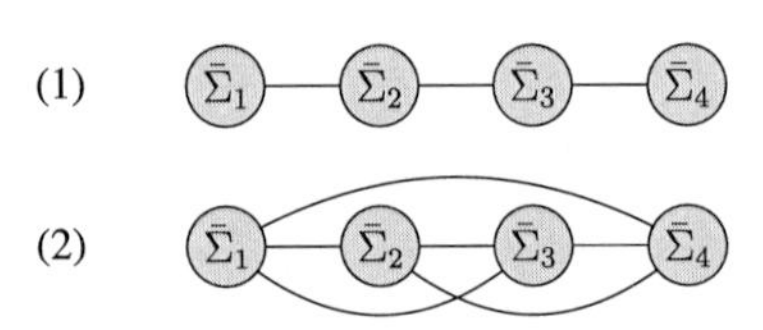

Figure 3.3: Interconnection structures: (1) chain-like coupling and (2) complete coupling

feedback

$$\boldsymbol{u}(t) = -\boldsymbol{K}_{\mathrm{OS}i}\,\boldsymbol{x}(t), \qquad i = 1, 2$$

is determined by solving the optimization problem (3.15) – (3.16) for the weighting matrices

$$\boldsymbol{Q} = \begin{pmatrix} 100 & 0 \\ 0 & 10 \end{pmatrix} \qquad \text{and} \qquad \boldsymbol{R} = 0.6.$$

For both interconnection structures, the resulting optimal feedback requires an information exchange between all mobile robots. In the case of the interconnection structure (1), the optimal full state feedback is given by

$$\boldsymbol{K}_{\mathrm{OS}1} = \begin{pmatrix} 10.5 & 12.3 & -7 & -7.4 & -2.1 & -2.9 & -1.4 & -2.1 \\ -7 & -7.4 & 15.5 & 16.8 & -6.3 & -6.5 & -2.1 & -2.9 \\ -2.1 & -2.9 & -6.3 & -6.5 & 15.5 & 16.8 & -7 & -7.4 \\ -1.4 & -2.1 & -2.1 & -2.9 & -7 & -7.4 & 10.5 & 12.3 \end{pmatrix}.$$

In the case of the complete agent coupling described by network (2), the optimal synchronizing feedback can be decomposed into local controllers

$$\boldsymbol{K}_2 = \begin{pmatrix} 6.5 & 6.7 \end{pmatrix}$$

and the communication network described by the Laplacian matrix

$$\boldsymbol{L}_{\mathcal{K}} = \begin{pmatrix} 3 & -1 & -1 & -1 \\ -1 & 3 & -1 & -1 \\ -1 & -1 & 3 & -1 \\ -1 & -1 & -1 & 3 \end{pmatrix}.$$

The resulting overall feedback is then given by $\boldsymbol{K}_{\mathrm{OS2}} = \boldsymbol{L}_{\mathcal{K}} \otimes \boldsymbol{K}_2$.

Figure 3.4 illustrates the experimental results in which asymptotic synchronization is confirmed for both network structures. Even if the feedback matrices $\boldsymbol{K}_{\mathrm{OS1}}$ and $\boldsymbol{K}_{\mathrm{OS2}}$ require

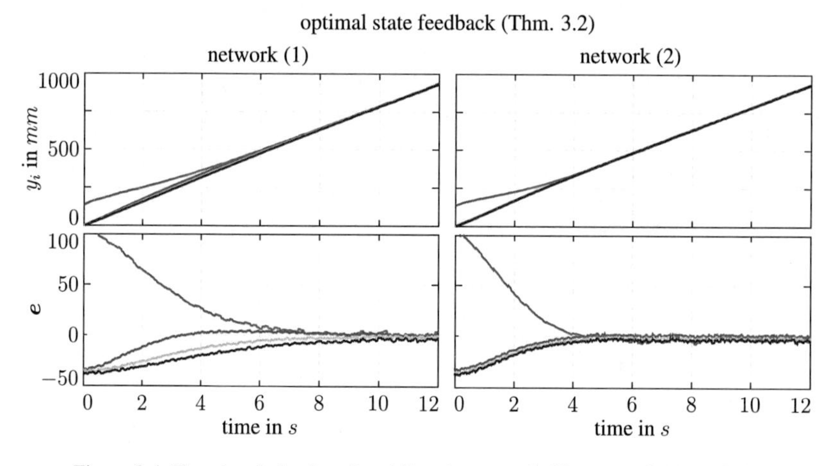

Figure 3.4: Transient behavior of mobile robots coupled by network (1) and (2)

complete state information, the behavior of the mobile robots which is observed is completely different. The differences can be explained by considering the state weighting term of the corresponding objective function in eqn. (3.15). Since the weighting is realized according to the network structure, the state difference between neighboring mobile robots is more relevant in network (1) as compared to network (2). The resulting controller $\boldsymbol{K}_{\mathrm{OS1}}$ produces a behavior, which is desired when satisfying the structural constraints that are provided by network (1).

That is, in the special case of a chain-like network the networked controller is only able to respond to a deviation of neighboring agents and from agent to agent. Hence, a delayed reaction of the mobile robots $i = 2, 3, 4$ on the deviation of the first mobile robot is observed.

In the case of network (2), all mobile robots have the same control input which produces a behavior where groups of synchronized agents ($i = 2, 3, 4$) remain synchronized among each other until all agents have asymptotically reached the synchronous trajectory. This phenomenon was considered in detail in [128].

The left side of Fig. 3.5 shows another experimental results for chain-like coupling structure and the design procedure presented in Theorem 3.2. In contrast to the behavior of the mobile robots shown in the left side of Fig. 3.4, the weighting of the control inputs is increased ($\boldsymbol{R} = 1$). A comparison of the experiments shows that reducing or increasing the scalar ratio between the state and the input weighting term directly reduces or increases the time for synchronization, respectively.

Approximation of the optimal synchronizing full state feedback

The main objective of the approximate networked controller presented in Section 3.8 is to transfer the optimal state feedback provided by Theorem 3.2 to a controller that satisfies the structural constraints of a networked controller as depicted in Fig. 3.1. First, the optimal state feedback $\boldsymbol{K}_{\mathrm{OS}}$ is obtained as the solution of the optimization problem (3.15) – (3.16) for the chain-like coupling structure and the weighting matrices

$$\boldsymbol{Q} = \begin{pmatrix} 100 & 0 \\ 0 & 10 \end{pmatrix} \qquad \text{and} \qquad \boldsymbol{R} = 1.$$

The corresponding synchronizing optimal feedback is given by

$$\boldsymbol{K}_{\mathrm{OS}} = \begin{pmatrix} 8.2 & 10 & -5.4 & -6.1 & -1.6 & -2.3 & -1.1 & -1.6 \\ -5.4 & -6.1 & 12 & 13.8 & -4.9 & -5.4 & -1.6 & -2.3 \\ -1.6 & -2.3 & -4.9 & -5.4 & 12 & 13.8 & -5.4 & -6.1 \\ -1.1 & -1.6 & -1.6 & -2.3 & -5.4 & -6.1 & 8.2 & 10 \end{pmatrix}$$

and the solutions of the algebraic riccati equations (3.24) by

$$\boldsymbol{P}_1 = \begin{pmatrix} 0 & 0 \\ 0 & 0 \end{pmatrix}, \quad \boldsymbol{P}_2 = \begin{pmatrix} 248 & 273 \\ 273 & 361 \end{pmatrix}, \quad \boldsymbol{P}_3 = \begin{pmatrix} 553 & 505 \\ 505 & 599 \end{pmatrix}, \quad \boldsymbol{P}_4 = \begin{pmatrix} 795 & 660 \\ 660 & 740 \end{pmatrix}.$$

Note that the spectrum of the Laplacian matrix

$$\boldsymbol{L}_{\mathcal{C}} = \begin{pmatrix} 1 & -1 & 0 & 0 \\ -1 & 2 & -1 & 0 \\ 0 & -1 & 2 & -1 \\ 0 & 0 & -1 & 1 \end{pmatrix}$$

of the communication network (1) is given by

$$\Lambda = \{0,\ 0.59,\ 2,\ 3.41\}. \tag{3.57}$$

According to Theorem 3.3, condition (3.33) has to be checked in order to determine the approximate networked controller with (3.34) – (3.35). From (3.57) it is easy to see that

$$\frac{1}{2}\left(\sum_{k=2}^{4} 1/\lambda_k^2\right)^{-1} \sum_{i=2}^{4} \frac{1}{\lambda_i} = 0.39$$

and $\lambda_2 \geq 0.39$ holds. Since condition (3.33) is fulfilled, the synchronizing networked controller

$$\tilde{\boldsymbol{K}}_{\mathrm{OS}} = \boldsymbol{L}_{\mathcal{C}} \otimes \boldsymbol{K}$$

with

$$\boldsymbol{K} = \boldsymbol{R}^{-1}\boldsymbol{B}^{T}\boldsymbol{X} = \begin{pmatrix} 11.6 & 15.1 \end{pmatrix}$$

is obtained as the solution of the minimization problem (3.32).

The experimental results presented in Fig. 3.5 show that the behavior of the mobile robots coupled by the approximate networked controller is close to the behavior of the optimal full state feedback. Although the output signals behave the same, there are differences in the required maximum control input amplitude, which is in the case of the approximate networked controller larger.

Optimal synchronizing networked controller

The design method for optimal networked controllers which is presented in Section 3.9 is considered for the chain-like coupling structure depicted in Fig. 3.3. First, an initial networked controller which satisfies the requirement on asymptotic synchronization is analyzed with respect to the overall system performance expressed by the objective function $\tilde{J}\left(\boldsymbol{K}_{\mathrm{init}}\right)$. Secondly, Algorithm 3.1 is applied to improve the initial controller $\boldsymbol{K}_{\mathrm{init}}$ to additionally satisfy the requirement on optimal synchronization.

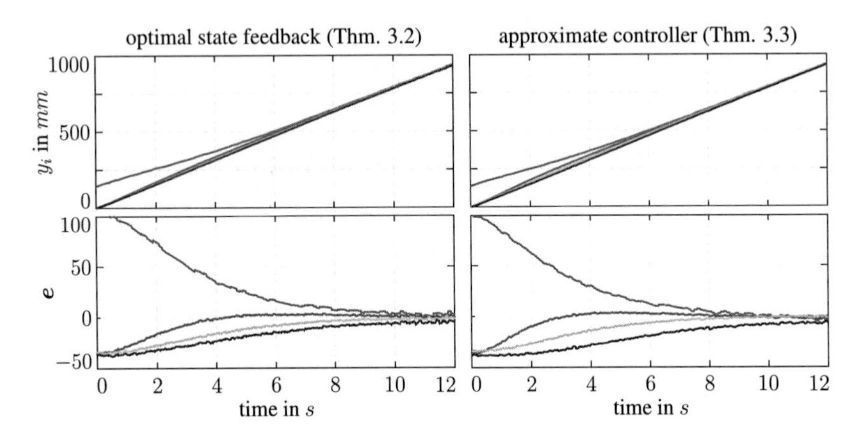

Figure 3.5: The optimal and the approximate synchronizing feedback

The parameters of the asymptotically synchronizing controller given by

$$\boldsymbol{K}_{\text{init}} = \begin{pmatrix} 8 & -5 \end{pmatrix}$$

are designed by a standard synchronizing controller design methods which only focuses on the synchronization condition (3.11). The left side of Fig. (3.6) shows the corresponding behavior of the mobile robots. The performance index of the initial controller is measured with respect to the quadratic cost given by

$$\tilde{J}\left(\boldsymbol{K}_{\text{init}}\right) = 14618.$$

The networked mobile robots obviously do not fulfill the requirement on optimal synchronization, which can be seen from the oscillations in Fig. 3.6 that are compared to the behavior shown in Fig. 3.5 too large.

The application of Algorithm 3.1 produces the control matrix

$$\boldsymbol{K}_{\text{opt}} = \begin{pmatrix} 7.1 & 8 \end{pmatrix}$$

which reveals a performance index

$$\tilde{J}\left(\boldsymbol{K}_{\text{opt}}\right) = 3468$$

that is four times smaller compared to the performance index of the initial networked controller. The right side of Fig. 3.6 shows that the behavior of the overall system has been

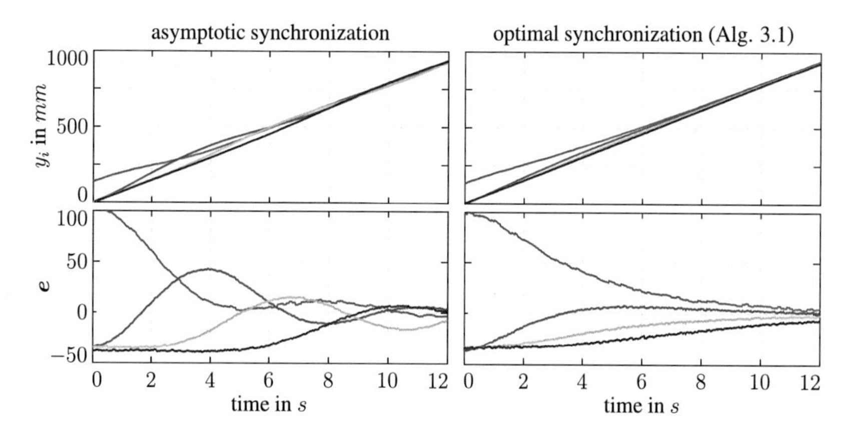

Figure 3.6: Transient behavior of mobile robots networked by an asymptotically and an optimal synchronizing networked controller

improved. Compared to the behavior of the approximate networked controller, the optimal networked controller leads to slower synchronization of the mobile robots with a lower maximum control input amplitude.

Publications

Optimal synchronization of agents coupled by a complete circulant structured communication network was considered in [1]. The general optimization problem of multi-agent systems coupled by a complete communication network and the corresponding approximate networked controller were considered in [2] and [4]. The solution of the extended optimization problem for the optimal synchronization of arbitrary coupled agents has been presented in [3].

4 Output synchronization with guaranteed performance

4.1 Introduction and performance requirements

This chapter presents an algorithm for the design of dynamic synchronizing controllers. If the states of the agents are communicated, it is sufficient to use networked controllers with static local components $\boldsymbol{K}$ to achieve synchronization with desired performance as shown in the previous chapter. If in contrast the output signals are communicated and in addition to the requirement on asymptotic synchronization temporal specifications of the transient behavior are made, it is generally necessary to use dynamic local controllers $\Sigma_{\mathrm{C}i}$, $(i = 1, 2, \ldots, N)$ as shown in Fig. 4.1.

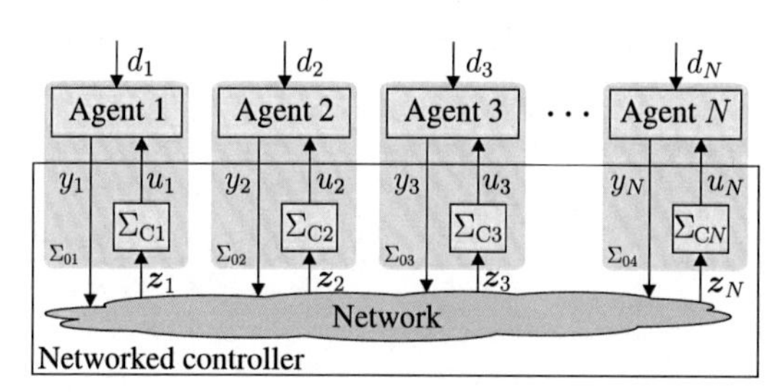

Figure 4.1: Networked multi-agent system

Synchronization of networked multi-agent systems is often considered for agents without output disturbances. Since in most practical applications there exist disturbances that affect the agents, asymptotic synchronization in the presence of step disturbances $d_i(t) = d_i\,\sigma(t)$ is considered. For all $d_i \in \mathbb{R}$, $(i = 1, 2, \ldots, N)$ the networked controller should ensure the convergence of all output signals $y_i(t)$ to a common trajectory $y_{\mathrm{s}}(t)$, which is expressed by

$$\lim_{t\to\infty} \|y_i(t) - y_{\mathrm{s}}(t)\| = 0, \qquad i = 1, 2, \ldots, N. \tag{4.1}$$

In addition to the requirement on *asymptotic synchronization*, this chapter addresses the transient behavior of the networked agents specified by the following two requirements:

(R1) *Synchronization time*: The agents should be synchronized within the five-percent settling time $T_{5\%}$:

$$\left(\sum_{i=1}^{N}\|y_i(t) - y_s(t)\|^2\right)^{\frac{1}{2}} \leq 0.05\,\|\boldsymbol{d}\|, \quad \text{for } t \geq T_{5\%}, \tag{4.2}$$

where $\boldsymbol{d} = (d_1, \ldots, d_N)^T$ is the stack of the disturbance amplitudes.

(R2) *Damping*: The transient behavior of the overall system should satisfy a given minimum damping.

The first part of this chapter presents a necessary and sufficient condition for synchronization of the networked multi-agent system. It is shown that asymptotic synchronization is achieved if a set of decoupled error dynamics results in a stable closed-loop behavior.

In the second part, the performance analysis and the controller design are studied under the assumption of identical integral first-order lag open-loop systems Σ_{0i} for undirected network structures. The result of the analysis shows that the requirement on the synchronization time can be guaranteed by selecting the controller parameters with respect to the smallest nonzero eigenvalue of the matrix $\boldsymbol{L}$ specifying the network. The additional requirement on a minimum damping is satisfied if the controller parameters are adjusted by means of the largest nonzero eigenvalue of $\boldsymbol{L}$.

According to the performance analysis, a new design method for dynamic networked controllers is developed. As an important issue, this design method is motivated by frequency domain and root locus methods to treat the transient behavior of the agents, whereas most methods found in the literature only consider synchronization as an asymptotic property. Therefore, a review of the root locus method is presented in Section 2.4 as this approach forms the basis for the controller design. The last part of this chapter presents extensions of the design approach involving the synchronization of multi-agent systems with non-identical dynamics and directed communication networks.

Structure of this chapter. Models of the agents, the dynamic networked controller and the overall multi-agent system are introduced in Section 4.2, Section 4.3 and Section 4.4, respectively. The synchronization condition and the corresponding decomposition of the networked multi-agent system into the synchronous behavior and the uncoupled error dynamics is presented in Section 4.5. The main result of this chapter is the controller design algorithm derived in Section 4.6. The design approach is extended to the synchronization of non-identical agents and directed interconnection structures in Section 4.7. Section 4.8 illustrates the design approach by application to the control of a vehicle platoon.

4.2 Agents with output disturbances

This chapter is devoted to the synchronization of N SISO systems with output disturbances, which are described by the state-space models

$$\Sigma_i : \begin{cases} \dot{\boldsymbol{x}}_i(t) = \boldsymbol{A}\,\boldsymbol{x}_i(t) + \boldsymbol{b}\,u_i(t), & \boldsymbol{x}_{0i} = \boldsymbol{0} \\ y_i(t) = \boldsymbol{c}^T\,\boldsymbol{x}_i(t) + d_i(t), \end{cases} \tag{4.3}$$

where

- $\boldsymbol{x}_i(t)$ is the n-dimensional state vector,
- $u_i(t)$ is the scalar control input,
- $d_i(t)$ is the scalar disturbance input and
- $y_i(t)$ is the scalar output of the i-th agent.

The class of disturbance signals is restricted to piecewise constant signals.

Assumption 4.1. *The agents are assumed to be affected by step output disturbances*

$$d_i(t) = d_i\,\sigma(t).$$

In order to avoid trivial solutions, agents with integral or unstable dynamics are considered.

Assumption 4.2. *The system matrix* $\boldsymbol{A}$ *of the state-space model* (4.3) *is assumed to has at least one eigenvalue with a non-negative real part:*

$$\exists\ \lambda_i(\boldsymbol{A}),\ i = 1, 2, \ldots, n\ :\ \Re(\lambda_i(\boldsymbol{A})) \geq 0.$$

Moreover, the agents are assumed to be completely controllable and completely observable.

Assumption 4.3. *The agents Σ_i are completely controllable*

$$\operatorname{rank}\begin{pmatrix}\boldsymbol{b} & \boldsymbol{A}\boldsymbol{b} & \boldsymbol{A}^2\boldsymbol{b} & \cdots & \boldsymbol{A}^{n-1}\boldsymbol{b}\end{pmatrix} = n$$

and completely observable:

$$\operatorname{rank}\begin{pmatrix}\boldsymbol{c}^T \\ \boldsymbol{c}^T\boldsymbol{A} \\ \boldsymbol{c}^T\boldsymbol{A}^2 \\ \vdots \\ \boldsymbol{c}^T\boldsymbol{A}^{n-1}\end{pmatrix} = n.$$

The representation of the agents Σ_i in the frequency domain is given by

$$Y_i(s) = G(s)\,U_i(s) + D_i(s), \qquad i = 1,\, 2,\, 3,\, \ldots,\, N$$

with $Y_i(s) \bullet\!\!-\!\!\circ\; y_i(t)$, $U_i(s) \bullet\!\!-\!\!\circ\; u_i(t)$, $D_i(s) \bullet\!\!-\!\!\circ\; d_i(t)$ and the strictly proper transfer function $G(s) = \boldsymbol{c}^T\left(s\boldsymbol{I} - \boldsymbol{A}\right)^{-1}\boldsymbol{b}$ describing the input-output behavior. Note that in virtue of Assumption (4.3) there is no pole-zero cancellation in the transfer function $G(s)$. Hence, all modes of the state-space models Σ_i are included in the input-output behavior.

The model of the overall multi-agent system is derived by the stack of output signals $Y_i(s)$:

$$\boldsymbol{Y}(s) = G(s)\,\boldsymbol{U}(s) + \boldsymbol{D}(s) \tag{4.4}$$

where

$$\begin{aligned}\boldsymbol{U}^T(s) &= (U_1(s),\, \ldots,\, U_N(s)),\\ \boldsymbol{Y}^T(s) &= (Y_1(s),\, \ldots,\, Y_N(s)),\\ \boldsymbol{D}^T(s) &= (D_1(s),\, \ldots,\, D_N(s)).\end{aligned}$$

According to Assumption 4.1, the disturbance signal $D(s)$ is specified by

$$\boldsymbol{D}(s) = \boldsymbol{d}\,\frac{1}{s} \tag{4.5}$$

with $\boldsymbol{d} = (d_1,\, \ldots,\, d_N)^T \in \mathbb{R}^N$.

4.3 Dynamic networked controller

Figure 4.1 shows the structure of the networked controller, which consists of the local controllers $\Sigma_{\mathrm{C}i}$ and the communication network. This chapter focuses on the derivation of a design method for the local controllers $\Sigma_{\mathrm{C}i}$ to satisfy (4.2) for a given communication network, which is assumed to exchange output signals between neighboring agents in the following way.

Assumption 4.4. *The communication network which is assumed to be given, provides the local controllers $\Sigma_{\mathrm{C}i}$ with the communicated coupling signals*

$$\begin{aligned} z_i(t) &= \sum_{j \in \mathcal{N}_i} (y_i(t) - y_j(t)), \qquad i = 1, 2, 3, \dots, N, \\ &= \sum_{j=1}^{N} l_{ij} (y_j(t) - y_i(t)) \end{aligned} \tag{4.6}$$

where $\mathcal{N}_i$ is the set of neighbors of agent Σ_i and l_{ij} the entries of a Laplacian matrix $\boldsymbol{L} = (l_{ij})$.

In Assumption 4.4 it is assumed that the structure of the communication network is given, which is important for the design method proposed in this chapter. The representation of the network structure is obtained by the sets $\mathcal{N}_i$, $(i = 1, 2, \dots, N)$ of neighbors or equally by using a communication graph which is expressed by a Laplacian matrix $\boldsymbol{L}$ of the form (2.3). The local controllers $\Sigma_{\mathrm{C}i}$ together with the communicated coupling signals (4.6) establish a diffusive coupling of the agents based on the exchange of the output signals $y_i(t)$. This chapter investigates the synthesis of networked controllers, where the control input of the i-th agent is calculated by the dynamic feedback

$$U_i(s) = -K(s)\, Z_i(s) \tag{4.7}$$

$$= -K(s) \sum_{j=1}^{N} l_{ij} (Y_i(s) - Y_j(s)) \tag{4.8}$$

with $Z_i(s) \bullet\!\!-\!\!\circ\, z_i(t)$. Under the common assumption of identical agent dynamics, there is no need for the use of non-identical local controllers $\Sigma_{\mathrm{C}i} = \Sigma_{\mathrm{C}}$. However, this situation changes in Section 4.7.1, where the synchronization of agents with non-identical dynamics is considered.

The interconnection structure of the overall system, which is a consequence of the coupling signals (4.6) is given by

$$\boldsymbol{Z}(s) = \boldsymbol{L}\,\boldsymbol{Y}(s) \tag{4.9}$$

with $\boldsymbol{Z}^T(s) = (Z_1(s), \ldots, Z_N(s))$. From eqn. (4.7) and (4.9) it is obvious that the networked controller consists of the local controllers $K(s)$ and the Laplacian matrix $\boldsymbol{L}$, which is associated to the communication network:

$$\boldsymbol{U}(s) = -K(s)\,\boldsymbol{L}\,\boldsymbol{Y}(s). \tag{4.10}$$

4.4 Model of the overall system and assumptions

The behavior of the overall closed-loop system with respect to the disturbance input $\boldsymbol{D}(s)$ is obtained from the agent model (4.4) and the networked controller (4.10):

$$\boldsymbol{Y}(s) = (\boldsymbol{I} + G(s)\,K(s)\,\boldsymbol{L})^{-1}\,\boldsymbol{D}(s). \tag{4.11}$$

The key idea for the design of the controllers $K(s)$ to satisfy the performance requirements on the closed-loop system (4.11) is based on the creation of a dominant pair of poles in the dynamics of decoupled synchronization errors introduced in Section 4.5. The following analysis shows that the transfer function of the closed-loop system reveals simple dynamics, which are suitable for the derivation of relations between the required performance specifications and the location of the dominant pair of poles. The following assumption involves the required simple dynamics of the networked agents.

Assumption 4.5. *It is assumed that there exists a controller $K(s)$ such that*

$$\frac{1}{1 + G(s)\,K(s)} \approx 1 - \frac{1}{\left(\dfrac{s}{\omega_0}\right)^2 + 2\,\dfrac{d}{\omega_0}\,s + 1} \tag{4.12}$$

or for the sake of simplicity

$$\frac{G(s)\,K(s)}{1 + G(s)\,K(s)} \approx \frac{1}{\left(\dfrac{s}{\omega_0}\right)^2 + 2\,\dfrac{d}{\omega_0}\,s + 1}$$

holds.

They are mainly two strategies that can be used for the design of the controller $K(s)$ to satisfy (4.12). First, the controller dynamics can be designed to cancel poles and zeros of the agent transfer function $G(s)$ in the open-loop systems $G_0(s) = G(s)\,K(s)$ shown in Fig. 4.1. Secondly, the approximation of the open-loop dynamics can also be attained by shaping the corresponding root locus without the cancellation of poles and zeros. The important difference between these strategies is that in the case of canceled poles and zeros the corresponding modes become unobservable in the open loop. Hence, initial conditions that are related to the canceled pole can not be compensated by the control loop. However, this drawback can be avoided by using the second strategy, which is also applied in the example in Section 4.8.

Assumption 4.5 is motivated by the control of a vehicle platoon where the vehicle dynamics are described by the transfer function

$$G(s) = \frac{1}{s\,(T_1\,s+1)}. \tag{4.13}$$

Note that in most applications, additional dynamics of vehicles exhibit significantly faster modes that can be neglected. Therefore, the simple integral first-order model represents a suitable approximation of many higher-order models of autonomous mobile systems. Assumption 4.5 with the agent model (4.13) is for example satisfied if the controller is designed as

$$K(s) = k\,\frac{T_1\,s+1}{T_{\mathrm{K}}\,s+1}.$$

In literature, synchronization of multi-agent systems is typically considered for nonzero initial conditions $\boldsymbol{x}_i(0) = \boldsymbol{x}_{0i}$, $(i = 1, 2, \ldots, N)$. This chapter deals in general with the synchronization of agents with output disturbances (4.3). But as it is shown in Fig. 4.2, nonzero initial values x_{0i} of integral dynamics and step disturbances have the same effect on the system's output. Hence, practical applications where the synchronization of e.g. the

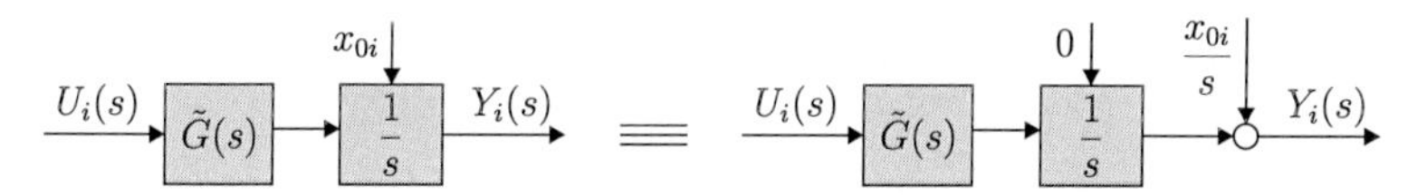

Figure 4.2: Effect of the initial value of the integrator on the output signal

vehicles (4.13) from different initial positions is of interest are also covered by the design procedure presented in this chapter.

4.5 Decomposition and synchronization condition

The following computations show that the transfer function matrix

$$\boldsymbol{S}(s) = (\boldsymbol{I} + G(s)\,K(s)\,\boldsymbol{L})^{-1} \tag{4.14}$$

describing the closed-loop multi-agent system can be decomposed into a set of uncoupled closed-loop systems of a subsystems order. The next lemma presents this well known decomposition, which is used for the controller design and the derivation of the synchronization condition.

Lemma 4.1 (Decomposition of the closed-loop dynamics)**.** *Let $\boldsymbol{T}$ be the eigenvector matrix of $\boldsymbol{L}$ with $\boldsymbol{T}^{-1}\boldsymbol{L}\,\boldsymbol{T} = \operatorname{diag}(\lambda_i)$. Then, the decomposition of $\boldsymbol{S}(s)$ is given by the diagonal transfer function matrix*

$$\tilde{\boldsymbol{S}}(s) = \boldsymbol{T}^{-1}\boldsymbol{S}(s)\boldsymbol{T} = \operatorname{diag}\left(\tilde{S}_i(s)\right), \tag{4.15}$$

where

$$\tilde{S}_i(s) = \frac{1}{1 + \lambda_i\,G(s)\,K(s)}, \quad i = 1, 2, \ldots, N.$$

Proof. The decomposition (4.15) results from the diagonizability of the Laplacian matrix $\boldsymbol{L}$:

$$\begin{aligned}
\tilde{\boldsymbol{S}}(s) &= \boldsymbol{T}^{-1}\,(\boldsymbol{I} + G(s)\,K(s)\,\boldsymbol{L})^{-1}\,\boldsymbol{T} \\
&= \left(\boldsymbol{I} + G(s)\,K(s)\,\boldsymbol{T}^{-1}\,\boldsymbol{L}\,\boldsymbol{T}\right)^{-1} \\
&= (\boldsymbol{I} + G(s)\,K(s)\,\operatorname{diag}(\lambda_i))^{-1} \\
&= \operatorname{diag}\left(\frac{1}{1 + \lambda_i\,G(s)\,K(s)}\right).
\end{aligned}$$

□

The first requirement to be fulfilled by the controlled multi-agent system is the requirement on asymptotic synchronization of the agents defined in eqn. (4.1). The following theorem gives conditions on asymptotic synchronization that are based on the analysis of the decoupled dynamics (4.15). Note that related results were first presented in [38].

Theorem 4.1 (Synchronization condition). *The networked multi-agent system (4.11) is asymptotically synchronized if and only if the following three aspects are fulfilled.*

1. *The open-loop transfer function $G_0(s)$ includes integrator dynamics,*
2. $\lambda_i > 0$, $(i = 2, 3, \ldots, N)$ *and*
3. *each pole of the decoupled dynamics*

$$\tilde{S}_i(s) = \frac{1}{1 + \lambda_i G(s) K(s)}, \quad i = 2, 3, \ldots, N \tag{4.16}$$

has a negative real part.

Proof. The details of the proof are given in the appendix. □

The first important result of Theorem 4.1 is that the eigenvalues λ_i, $(i = 2, 3, \ldots, N)$ of the Laplacian matrix $\boldsymbol{L}$ have to be different from zero in order to achieve asymptotic synchronization of the networked multi-agent system (4.11). From the algebraic graph theory it is well known that the second largest Laplacian eigenvalue λ_2 is closely related to connectivity properties of the underlying communication graph, namely, the communication graph has a spanning tree if and only if $\lambda_2 > 0$. Therefore, the following is assumed.

Assumption 4.6. *The communication network which is associated to the Laplacian matrix* $\boldsymbol{L}$ *is assumed to be connected* $(\lambda_2 > 0)$.

Secondly, Theorem 4.1 shows that the part of the decoupled dynamics (4.15) belonging to the non-zero eigenvalues of the Laplacian matrix $\boldsymbol{L}$ is critical for asymptotic synchronization of the agents, since it is necessary that the decoupled dynamics in (4.16) are input-output stable. Therefore, the decoupled dynamics in (4.16) are in what follows referred to as the *decoupled synchronization error dynamics*. The following lemma combines the results of Lemma 4.1 and Theorem 4.1 and presents a decomposition of the overall multi-agent system (4.11) into the behavior of the synchronization error and the synchronous trajectory.

Lemma 4.2 (Decomposition of the agent behavior). *The output signal $\boldsymbol{Y}(s)$ of the overall multi-agent system (4.11) is composed of the synchronous trajectory $\boldsymbol{Y}_{\mathrm{s}}(s)$ and the synchronization error $\boldsymbol{E}(s)$:*

$$\boldsymbol{Y}(s) = \boldsymbol{Y}_{\mathrm{s}}(s) + \boldsymbol{E}(s), \tag{4.17}$$

where

$$\boldsymbol{Y}_{\mathrm{s}}(s) = \begin{pmatrix} Y_{\mathrm{s}}(s) & Y_{\mathrm{s}}(s) & \cdots & Y_{\mathrm{s}}(s) \end{pmatrix}^T, \quad Y_{\mathrm{s}}(s) = \frac{1}{N}\sum_{i=1}^{N} d_i \frac{1}{s} \tag{4.18}$$

and

$$\boldsymbol{E}(s) = \boldsymbol{T} \begin{pmatrix} 0 & & & \\ & \tilde{S}_2(s) & & \\ & & \ddots & \\ & & & \tilde{S}_N(s) \end{pmatrix} \boldsymbol{T}^{-1}\boldsymbol{d}\frac{1}{s}. \tag{4.19}$$

Proof. The proof of the lemma results from the transformation of the overall system (4.11) by means of the eigenvector matrix $\boldsymbol{T}$ given by

$$\begin{aligned} \boldsymbol{Y}(s) &= \boldsymbol{T}\,\tilde{\boldsymbol{S}}(s)\,\boldsymbol{T}^{-1}\,\boldsymbol{D}(s) \\ &= \underbrace{\boldsymbol{T}\,\tilde{\boldsymbol{S}}_{\mathrm{s}}\boldsymbol{T}^{-1}\,\boldsymbol{D}(s)}_{\boldsymbol{Y}_{\mathrm{s}}(s)} + \underbrace{\boldsymbol{T}\,\tilde{\boldsymbol{S}}_{\Delta}(s)\,\boldsymbol{T}^{-1}\,\boldsymbol{D}(s)}_{\boldsymbol{E}(s)} \end{aligned} \tag{4.20}$$

with

$$\tilde{\boldsymbol{S}}(s) = \underbrace{\begin{pmatrix} 1 & & & \\ & 0 & & \\ & & \ddots & \\ & & & 0 \end{pmatrix}}_{\tilde{\boldsymbol{S}}_{\mathrm{s}}} + \underbrace{\begin{pmatrix} 0 & & & \\ & \tilde{S}_2(s) & & \\ & & \ddots & \\ & & & \tilde{S}_N(s) \end{pmatrix}}_{\tilde{\boldsymbol{S}}_{\Delta}(s)}.$$

From Theorem 4.1 it is easy to see that $\lim_{s\to 0} \tilde{S}_i(s) = 0$, $(i = 2, 3, \ldots, N)$ and $\lim_{s\to 0} \boldsymbol{E}(s) = 0$ for the step disturbances (4.5) holds. Hence, the synchronization error $\boldsymbol{e}(t)$ vanishes asymptotically under the presence of step disturbances. The first term of the sum in (4.17) is calculated by considering eqn. (4.5), the orthogonality of $\boldsymbol{T}$ and by considering the first eigenvector of the Laplacian matrix which is given by $\boldsymbol{t}_1 = \mathbb{1}/\sqrt{N}$, $(\boldsymbol{T} = (\,\boldsymbol{t}_1 \;\boldsymbol{t}_2 \;\cdots\; \boldsymbol{t}_N\,))$:

$$\begin{aligned}\boldsymbol{Y}_{\mathrm{s}}(s) &= \boldsymbol{T}\,\tilde{\boldsymbol{S}}_{\mathrm{s}}\,\boldsymbol{T}^{-1}\,\boldsymbol{D}(s)\\ &= \boldsymbol{t}_1\boldsymbol{t}_1^T\boldsymbol{d}\,\frac{1}{s}\\ &= \mathbb{1}\frac{1}{N}\sum_{i=1}^{N} d_i\,\frac{1}{s}.\end{aligned}$$

Since, $\lim_{s\to 0}\boldsymbol{E}(s) = 0$ it follows directly from the decomposition (4.20) and the final value theorem that $\lim_{s\to 0}\boldsymbol{Y}(s) - \boldsymbol{Y}_{\mathrm{s}}(s) = \boldsymbol{0}$ and $\lim_{t\to\infty} y_i(t) - y_{\mathrm{s}}(t) = 0$, $(i = 1, 2, \ldots, N)$ with

$$y_{\mathrm{s}}(t) \circ\!\!-\!\!\bullet \frac{1}{N}\sum_{i=1}^{N} d_i\,\frac{1}{s},$$

which proves the lemma. □

The transformation of eqn. (4.18) into the time-domain shows that the synchronous trajectory is given by the mean value of the weighted step disturbances given by

$$Y_{\mathrm{s}}(s) \bullet\!\!-\!\!\circ\; y_{\mathrm{s}}(t) = \frac{1}{N}\sum_{i=1}^{N} d_i\,\sigma(t).$$

4.6 Design of dynamic networked controllers

4.6.1 Design idea

The proposed design method is inspired by well-known frequency domain methods that are used for the design of controllers for SISO systems. As described in Section 2.4, a controller that fulfills the requirements on the settling time $T_{5\%}$ and the overshoot peak Δh of the closed-loop system can be obtained in two steps. First the dynamics of the controller have to be designed such that the closed-loop system has approximately second-order dynamics which are characterized by the natural frequency ω_0 and the damping ratio d. Secondly, the relation between the performance requirements $(T_{5\%}, \Delta h)$ and the closed-loop characteristics (d, ω_0) has to be used for the determination of the controller parameters.

The same procedure is used for the derivation of the controller design method that fulfills the performance requirements listed in Section 4.1. In virtue of Assumption 4.5 there exist local controllers $K(s)$ that cause a behavior of the decoupled synchronization error dynamics (4.1), which can approximately be described by a second-order transfer function of the form (4.12). Note, that from the decomposition in Lemma 4.2 it is obvious that the synchronization error $\boldsymbol{E}(s)$, which is primarily influenced by the decoupled error dynamics, is given by a superposition of this simple second-order dynamics.

An important property of the decoupled error dynamics $\tilde{S}_i(s)$, $(i = 2, 3, \ldots, N)$ can be seen by considering their transfer functions as control loops that have the same open-loop function $G_0(s)$ but different open-loop gains λ_i (Fig. 4.3). According to this property and the fact

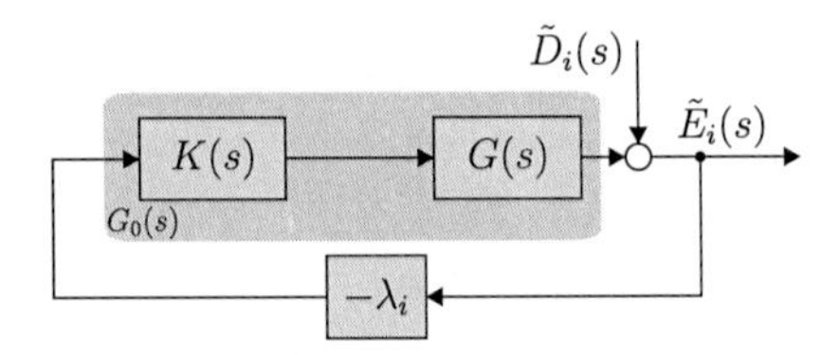

Figure 4.3: Structure of the transfer functions of the decoupled synchronization errors $\tilde{S}_i(s)$

that the decoupled synchronization errors are described by second-order transfer functions, it is reasonable that the transient behavior of decoupled synchronization errors is bounded with respect to the smallest λ_i, $(i = 2, 3, \ldots, N)$.

The main idea of the controller design procedure is to bring this bound into relation with the synchronization error $\boldsymbol{E}(s)$ and to translate the requirements on the settling time and the minimum damping into specifications of the upper bound, which are useful for the determination of the controller parameters.

In summary, the controller design method is obtained in the following three steps:

1. The relation between the decoupled synchronization errors $\tilde{S}_i(s)$, $(i = 2, 3, \ldots, N)$, the eigenvalues λ_i of the Laplacian matrix $\boldsymbol{L}$ and the time-domain behavior of the overall multi-agent system is examined.

2. The transient behavior of the decoupled error dynamics $\tilde{S}_i(s)$ is investigated and it is shown that the synchronization error $\boldsymbol{e}(t) \circ\!\!-\!\!\bullet\ \boldsymbol{E}(s)$ is bounded with respect to the second largest eigenvalue of the Laplacian matrix $\boldsymbol{L}$.

3. In the last step, the upper bound of the synchronization error $\boldsymbol{e}(t)$ is used for the determination of the controller parameters which ensure the requirements on the synchronization time and the minimum damping.

4.6.2 Transient behavior of the decoupled synchronization errors

The decoupled error dynamics $\tilde{S}_i(s)$, $(i = 2, 3, \ldots, N)$, which are defined in Theorem 4.1 can be interpreted as a feedback of the open-loop transfer function $G_0(s)$ that are weighted by the nonzero eigenvalues of the Laplacian matrix $\boldsymbol{L}$ (Fig. 4.3). Note that in the case of a bidirectional communication network, the Laplacian matrix is symmetric and has consequently only positive real-valued eigenvalues. According to Assumption 4.5, the decoupled error dynamics $\tilde{S}_i(s)$ are required to be approximately described by

$$\begin{aligned}\tilde{S}_i(s) &= \frac{1}{1+\lambda_i\, G_0(s)}, \qquad\qquad i = 2, 3, \ldots, N\\ &= 1 - \frac{\lambda_i\, G_0(s)}{1+\lambda_i\, G_0(s)}\\ &= 1 - \frac{1}{\left(\dfrac{s}{\omega_{0i}}\right)^2 + 2\,\dfrac{d_i}{\omega_{0i}}\,s + 1}\end{aligned} \tag{4.21}$$

where

$$\omega_{0i} = \sqrt{\lambda_i}\,\omega_0, \qquad d_i = \frac{d}{\sqrt{\lambda_i}} \tag{4.22}$$

and

$$G_0(s) = \frac{1}{\dfrac{s}{\omega_0}\left(\dfrac{s}{\omega_0} + 2\,d\right)} \tag{4.23}$$

holds.

In Section 2.4, it is stated that the open-loop gain of an integral first-order system determines the location of the closed-loop poles on the two branches of the root locus plot. Therefore, it is straightforward to analyze the location of the pole-pairs of $\tilde{S}_i(s)$, $(i = 2, 3, \ldots, N)$ in relation to the eigenvalues λ_i of the Laplacian matrix by considering the the results of Section 2.4. For the calculation of the poles of the decoupled error dynamics, the parametrized form in eqn. (4.21) is evaluated with eqn. (4.22) and the general description given in eqn. (2.19):

$$s_{1i,2i} = -\omega_{0i}\, d_i \pm \omega_{0i}\sqrt{d_i^2 - 1}, \quad i = 2, \ldots, N \tag{4.24}$$

$$= -\omega_0\, d \pm \omega_0\,\sqrt{d^2 - \lambda_i}. \tag{4.25}$$

Since the transfer functions $\tilde{S}_i(s)$, $(i = 2, 3, \ldots, N)$ have the form of the closed-loop transfer function (2.23), the corresponding step responses are described in the time domain by (2.24). The substitution of the parametrized natural frequencies ω_{0i}, the dampings d_i into eqn. (2.24) gives the transient responses $h_i(t) \circ\!\!-\!\!\bullet \tilde{S}_i(s)\frac{1}{s}$, $(i = 2, 3, \ldots, N)$:

$$h_i(t) = \begin{cases} \frac{\mathrm{e}^{-\omega_0\, d\, t}}{\sqrt{1-d_i^2}} \sin\left(\omega_0\sqrt{\lambda_i - d^2}\, t + \arccos(d_i)\right), & d_i < 1 \\ (1+\omega_{0i}\, t)\ \mathrm{e}^{-\omega_{0i}\, t}, & d_i = 1 \\ \frac{1}{s_{1i} - s_{2i}} \left(-s_{2i}\, \mathrm{e}^{s_{1i}\, t} + s_{1i}\, \mathrm{e}^{s_{2i}\, t}\right), & d_i > 1 \end{cases}. \tag{4.26}$$

Depending on the given network $\boldsymbol{L}$ and the open-loop $G_0(s)$, the set of the decoupled synchronization error dynamics with solutions $h_i(t)$ consists of elements that can exhibit underdamped $(d_i < 1)$, critically damped $(d_i = 1)$ and overdamped $(d_i > 1)$ characteristics. The relation between the transient responses $h_i(t)$, $(i = 2, 3, \ldots, N)$ and the settling time $T_{5\%}$ can, as described in Section 2.4 be concluded from the underdamped case in (4.26):

$$T_{5\%} \approx \frac{3 - \ln\left(\sqrt{1-d_i^2}\right)}{\omega_{0i}\, d_i}. \tag{4.27}$$

4.6.3 Bound on the overall synchronization error

Example 4.1 shows that the pole pairs belonging to the smallest nonzero Laplacian eigenvalue (λ_2) are dominant compared to the other poles. Larger Laplacian eigenvalues lead to an accelerated transient behavior of the decoupled error dynamics, which is obvious in the case of open-loop transfer functions with integral-first order dynamics. Hence, it is worth to consider the characteristic behavior of the transient responses $h_i(t)$, $(i = 2, 3, \ldots, N)$ with respect to the corresponding eigenvalues of the Laplacian matrix $\boldsymbol{L}$ in detail.

Example 4.1 *Transient behavior of decoupled error dynamics*

The following example considers the transient behavior of the decoupled error dynamics resulting from an interconnection of $N = 8$ agents. Figure 4.4 shows the structure of the communication network and the root locus plot of the open-loop transfer function

$$G_0(s) = \frac{0.109}{s\,(s+1)} \tag{4.28}$$

where the closed-loop poles belonging to the corresponding $N - 1 = 7$ nonzero eigenvalues

$$\lambda_2 = 0.38, \quad \lambda_3 = 0.73, \quad \lambda_4 = 1, \quad \lambda_5 = 2.3, \quad \lambda_6 = 2.62, \quad \lambda_7 = 3.51, \quad \lambda_8 = 5.46$$

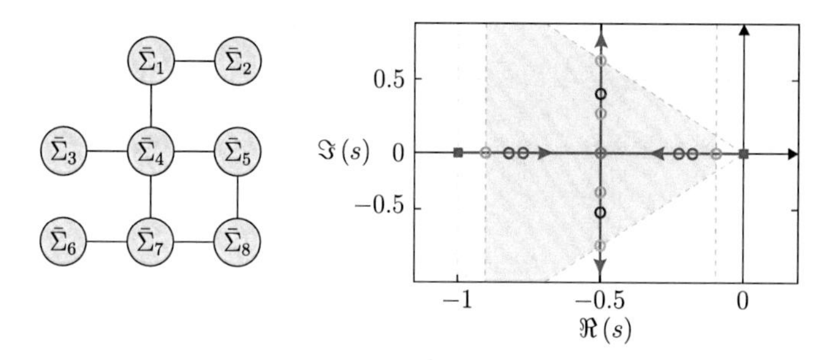

Figure 4.4: Example for the location of the poles of $\tilde{S}_i(s)$ for $N = 8$ agents

of the Laplacian matrix

$$
\boldsymbol{L} = \begin{pmatrix} 2 & -1 & 0 & -1 & 0 & 0 & 0 & 0 \\ -1 & 1 & 0 & 0 & 0 & 0 & 0 & 0 \\ 0 & 0 & 1 & -1 & 0 & 0 & 0 & 0 \\ -1 & 0 & -1 & 4 & -1 & 0 & -1 & 0 \\ 0 & 0 & 0 & -1 & 2 & 0 & 0 & -1 \\ 0 & 0 & 0 & 0 & 0 & 1 & -1 & 0 \\ 0 & 0 & 0 & -1 & 0 & -1 & 3 & -1 \\ 0 & 0 & 0 & 0 & -1 & 0 & -1 & 2 \end{pmatrix}
$$

are marked by circles.

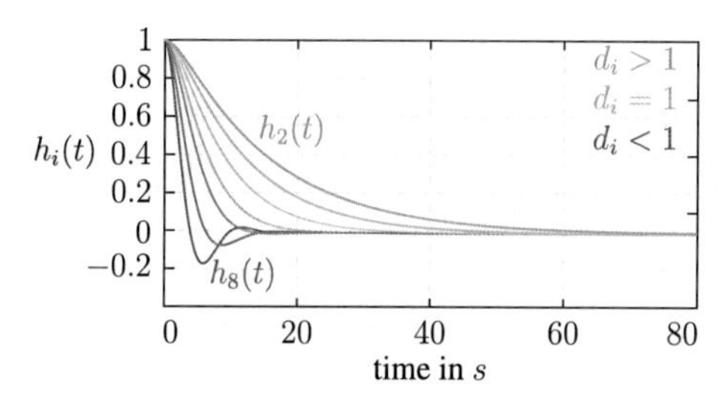

Figure 4.5: Example for the transient behavior of underdamped, critically damped and overdamped functions $h_i(t)$, $(i = 2, 3, \ldots, N)$

The decoupled error dynamics exhibit a characteristic behavior which is strongly related to the results presented in Section 2.4. Furthermore, Figure 4.4 reveals the existence of a dominant pair of poles belonging to the second Laplacian eigenvalue λ_2. Of particular interest is the link between the transient behavior corresponding to the dominant pair of poles on the *transient behavior of the remaining decoupled error dynamics. Figure 4.5 shows the transient responses $h_i(t)$, $(i = 2, 3, \ldots, 8)$ belonging to the open-loop transfer function (4.28) and the communication graph depicted in Fig. 4.4. Indeed, the transient response $h_2(t)$ is dominant in the sense that it posses the slowest convergence rate.*

The simulation in Example 4.1 reveals that the overdamped step responses are bounded by each other in magnitude. In the case of the underdamped transient responses this property is not fulfilled because of the oscillating behavior. However, it is evident for $d_i < 1$ that the amplitudes of the oscillations decrease within a common envelope.

The following lemma shows that this observation is generally valid and that the transient responses $h_i(t)$, $(i = 2, 3, \ldots, N)$ are bounded with respect to λ_2.

Lemma 4.3 (Magnitude relations of the transient responses $h_i(t)$)**.** *The transient responses $h_i(t)$ bound each other for $i = 2, 3, \ldots, N$:*

$$h_{i+1}(t) \leq h_i(t), \quad d_i \geq 1 \tag{4.29}$$

$$h_{i+1}^{d_{i+1}<1}(t) < h_i^{d_i>1}(t), \tag{4.30}$$

$$\hat{h}_{i+1}(t) \leq \hat{h}_i(t), \quad d_i < 1, \tag{4.31}$$

where

$$\hat{h}_i(t) = \frac{1}{\sqrt{1 - d_i^2}} \, \mathrm{e}^{-\omega_0 \, d \, t}$$

are the envelopes of the underdamped transient responses in (4.26).

Proof. The proof is given in the appendix. □

Lemma 4.3 provides a bound on the transient responses $h_i(t)$, $(i = 2, 3, \ldots, N)$ but not on the synchronization error $\boldsymbol{e}(t)$ which is important for meeting the requirements specified in Section 4.1. The result of Lemma 4.3 is applied in the following theorem to characterize the behavior of the overall synchronization error $\boldsymbol{e}(t) \circ\!\!-\!\!\bullet \boldsymbol{E}(s)$ defined in (4.19).

Theorem 4.2 (Bound on the synchronization error). *The Euclidean norm of the overall synchronization error $\mathbf{e}(t)$ is bounded by*

$$\|\boldsymbol{e}(t)\| \leq \begin{cases} |h_2(t)| \, \|\boldsymbol{d}\| \,, & d_2 \geq 1 \\ \left|\hat{h}_2(t)\right| \, \|\boldsymbol{d}\| \,, & d_2 < 1 \end{cases} \tag{4.32}$$

where $\mathbf{d} = (d_1, \ldots, d_N)^T$ describes the magnitude of the disturbance defined in (4.5).

Proof. Since, $\boldsymbol{e}(t) \circ\!\!-\!\!\bullet\, \boldsymbol{E}(s) = \boldsymbol{T}\,\tilde{\boldsymbol{S}}_\Delta(s)\boldsymbol{T}^{-1}\boldsymbol{d}\,\frac{1}{s}$, the norm of the synchronization error can be estimated by

$$\begin{aligned} \|\boldsymbol{e}(t)\| &= \left\|\boldsymbol{T}\,\boldsymbol{H}(t)\,\boldsymbol{T}^{-1}\,\boldsymbol{d}\right\| \\ &\leq \left\|\boldsymbol{T}\,\boldsymbol{H}(t)\,\boldsymbol{T}^{-1}\right\| \, \|\boldsymbol{d}\| \,, \end{aligned} \tag{4.33}$$

where

$$\boldsymbol{H}(t) = \begin{pmatrix} 0 & & & \\ & h_2(t) & & \\ & & \ddots & \\ & & & h_N(t) \end{pmatrix} \tag{4.34}$$

and $h_i(t) \circ\!\!-\!\!\bullet\, \tilde{S}_i(s)\frac{1}{s}$. Consider the spectral norm of matrices, then it is obvious that for orthogonal matrices $\boldsymbol{T}^T = \boldsymbol{T}^{-1}$,

$$\begin{aligned} \left\|\boldsymbol{T}\,\boldsymbol{H}(t)\,\boldsymbol{T}^{-1}\right\| &= \sqrt{\max_i \, \lambda_i \left(\boldsymbol{T}\,\boldsymbol{H}^2(t)\,\boldsymbol{T}^{-1}\right)} \\ &= \sqrt{\max_i \, h_i^2(t)} \\ &= |h_2(t)| \end{aligned}$$

holds. The proof follows in consideration of Lemma 4.3. □

The theorem is interesting because it provides an upper bound on the synchronization error $\boldsymbol{e}(t)$, which solely depends upon the scalar transient response $h_2(t)$. Note that $h_2(t)$ is related to the smallest eigenvalue λ_2 of the Laplacian matrix $\boldsymbol{L}$. From literature it is well known that the second largest Laplacian eigenvalue is important for the synchronizability of networked agents. Namely, synchronization of the agents can only be achieved if the communication graph is connected, which is the purpose of Assumption 4.6. However, Theorem 4.2 shows that the eigenvalue λ_2 is not only important for asymptotic synchronization but also for the transient behavior since it determines an upper bound that is useful for the design of the local controllers $K(s)$ to satisfy the requirements formulated in Section 4.1.

4.6.4 Controller design algorithm

The main result of this chapter is an algorithm for the design of dynamic local controller that satisfies the performance requirements (R1)–(R2). In the first step the requirement on the synchronization time is considered, as it can be commonly applied to all elements of the synchronization error $\boldsymbol{e}(t)$ in virtue of Theorem 4.2. The following corollary clarifies the relation between the upper bound (4.32) and the required synchronization time $T_{5\%}$.

Corollary 4.1 (Relation between the bound and the synchronization time). *The networked multi-agent system* (4.11) *is synchronized with respect to requirement* (4.2) *if the dynamics of the local controllers* $K(s)$ *are designed such that Assumption 4.5 is fulfilled and for* $t \geq T_{5\%}$ *and* $d_2 \geq 1$

$$|h_2(t)| \leq 0.05$$

or in the case that $d_2 < 1$

$$|\hat{h}_2(t)| \leq 0.05$$

holds.

Proof. The overall representation of the requirement (4.2) is obtained by

$$\left(\sum_{i=1}^{N} \|y_1(t) - y_s(t)\|^2\right)^{\frac{1}{2}} = \left\| \begin{matrix} y_1(t) - y_s(t) \\ y_2(t) - y_s(t) \\ \vdots \\ y_N(t) - y_s(t) \end{matrix} \right\| = \|\boldsymbol{e}(t)\|.$$

The combination of the requirement (4.2) and the result given in Theorem 4.2 gives the inequality

$$\|\boldsymbol{e}(t)\| \leq \begin{cases} |h_2(t)| \, \|\boldsymbol{d}\|, & d_2 \geq 1 \\ \left|\hat{h}_2(t)\right| \, \|\boldsymbol{d}\|, & d_2 < 1 \end{cases} \leq 0.05 \, \|\boldsymbol{d}\|,$$

which is fulfilled if

$$\left. \begin{array}{ll} |h_2(t)|\,, & d_2 \geq 1 \\ \left|\hat{h}_2(t)\right|\,, & d_2 < 1 \end{array} \right\} \leq 0.05$$

holds. □

Corollary 4.1 with eqns. (4.26), (4.27) and (4.22) shows that the settling time can either be set by the natural frequency ω_{02} or the damping d_2. Hence, specifying the settling time of the bound $h_2(t)$ solely employs one of the two available degrees of freedom. This fact can also be seen by considering Fig. 4.4, where the requirement on the settling time can be achieved by placing the poles of the decoupled dynamics $\tilde{S}_i(s)$ on the negative real axis as well as by choosing them to be strictly conjugate-complex or as a combination of both. Clearly, these different parametrizations have a significant impact on the transient behavior of the overall system. Therefore, it is reasonable to use the damping as a second tunable parameter besides the specified settling time to allow for an adjustment of the smoothness of the transient response. While the requirement on the settling time $T_{5\%}$ is associated with the bound $h_2(t)$, eqn. (4.22) shows that the requirement on a minimum damping d_{min} is related to the largest eigenvalue λ_N of the Laplacian matrix $\boldsymbol{L}$:

$$d_N = d_{\text{min}}. \tag{4.35}$$

Due to (4.22), the relation between the damping d_2 of the bound $h_2(t)$ and (4.35) is fixed by

$$d_2 = \sqrt{\frac{\lambda_N}{\lambda_2}}\, d_{\text{min}}. \tag{4.36}$$

Equation (4.36) allows to translate the requirement on the minimum damping into the requirement on the damping d_2. In virtue of Corollary 4.1 and eqn. (4.36), the design of the local controllers $K(s)$ that satisfies the performance requirements (R1) – (R2) is performed by specifying the natural frequency ω_{02} and the damping d_2 of the upper bound $h_2(t)$, which has the order of a single agent. To achieve this goal, the requirements $(d_2, T_{5\%})$ on the bound $h_2(t)$ are used for the derivation of the parameters T_{K} and k of the desired open-loop transfer function

$$G_0(s) = G(s)\,K(s) = \frac{k}{s\,(T_{\text{K}}\,s + 1)} \tag{4.37}$$

with

$$\omega_0 = \sqrt{\frac{k}{T_{\text{K}}}} \tag{4.38}$$

and

$$d = \frac{1}{2}\sqrt{\frac{1}{k\,T_{\mathrm{K}}}}. \tag{4.39}$$

In the first step the control gain k is obtained by substitution of (4.36) and (4.39) into (4.22) for $i = 2$:

$$k = \frac{\omega_{\mathrm{K}}}{4\,\lambda_N\,d_{\min}^2} \tag{4.40}$$

where

$$\omega_{\mathrm{K}} = \frac{1}{T_{\mathrm{K}}}.$$

The following three cases, which depend on the value of the required damping (4.36), have to be considered for the derivation of the frequency ω_{K}.

Case 1 ($d_2 < 1$)

For the underdamped case, the relation between the settling time $T_{5\%}$, the natural frequency ω_{02} and the damping d_2 is described by eqn. (4.27). Combining (4.22) for $i = 2$ with (4.36) and (4.27) gives the frequency ω_{K} as

$$\omega_{\mathrm{K}} \approx \frac{2}{T_{5\%}}\left(3 - \ln\left(\sqrt{1 - \frac{\lambda_N}{\lambda_2}\,d_{\min}^2}\right)\right). \tag{4.41}$$

Case 2 ($d_2 \geq 2$)

For $d_2 \geq 2$, the poles s_{12} and s_{22} defined by (4.25) show a significant distance on the negative real axis:

$$|s_{12}| < |s_{22}|.$$

Thus, the rightmost pole s_{12} dominates the transient response (4.26):

$$h_2(t) \approx \frac{-s_{22}}{s_{12} - s_{22}}\,\mathrm{e}^{s_{12}\,t} \approx \mathrm{e}^{s_{12}\,t}. \tag{4.42}$$

The relation between the settling time $T_{5\%}$ and the pole s_{12} is consequently obtained from (4.42):

$$0.05 \approx \mathrm{e}^{s_{12}\,T_{5\%}} \quad \Leftrightarrow \quad T_{5\%} \approx -\frac{3}{s_{12}}. \tag{4.43}$$

Substituting eqn. (4.22) into (4.25) yields

$$s_{12} = -\omega_0\, d + \sqrt{(\omega_0\, d)^2 - \omega_0^2 \lambda_2}$$
$$= -\frac{\omega_K}{2} + \sqrt{\left(\frac{\omega_K}{2}\right)^2 - \omega_K\, k\, \lambda_2} \tag{4.44}$$

and by considering eqn. (4.40)

$$\omega_K\, k\, \lambda_2 = \left(\frac{\omega_K}{2}\right)^2 \frac{\lambda_2}{\lambda_N\, d_{\min}^2}.$$

Hence, eqn. (4.44) can be rewritten as

$$s_{12} = \frac{\omega_K}{2}\left(-1 + \sqrt{1 - \frac{\lambda_2}{\lambda_N\, d_{\min}^2}}\right).$$

The relation between the real pole s_{12} and the settling time in (4.43) gives the design formula for the case $d_2 \geq 2$:

$$\omega_K \approx \frac{6}{T_{5\%}}\left(1 - \sqrt{1 - \frac{\lambda_2}{\lambda_N\, d_{\min}^2}}\right)^{-1}. \tag{4.45}$$

Case 3 $(1 \leq d_2 < 2)$

In this case, no analytical relation between the specified settling time $T_{5\%}$ and the poles s_{12} and s_{22} can be derived. In order to accomplish the controller design, the poles of the overdamped

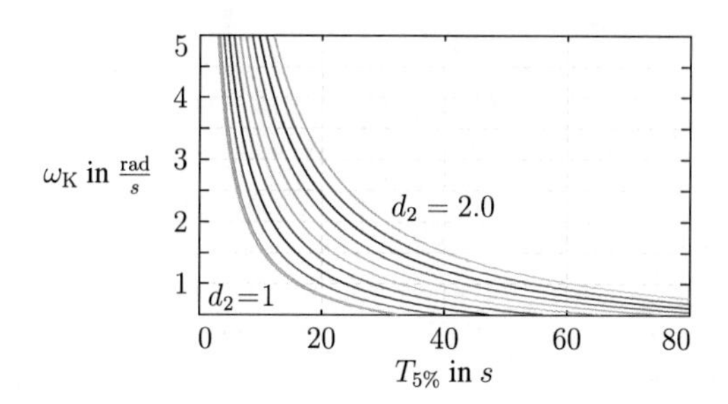

Figure 4.6: Relations between ω_K and $T_{5\%}$ for $1 \leq d_2 < 2$

case in (4.26) can be parametrized by ω_{K} and d_2 with (4.45) and (4.40):

$$s_{12,22} = \frac{\omega_{\mathrm{K}}}{2}\left(-1 \pm \sqrt{1 - \frac{1}{d_2^2}}\right). \tag{4.46}$$

If the parametrized poles (4.46) are inserted into the transient behavior (4.26) for the over-damped case, then the equation can be numerically solved for ω_{K} by considering the relation $h_2(T_{5\%}) = 0.05$. The required settling time $T_{5\%}$ can be directly respected where the damping d_2 has to be determined from the specified minimum damping d_{min} with (4.36). Figure 4.6 presents the resulting diagram to select ω_{K} for the required settling time $T_{5\%}$ and damping d_2.

Finally, the procedure for the design of the local controllers $K(s)$ is summarized in the following algorithm.

Algorithm 4.1. *Dynamic controller design*

Given: *$G(s)$, $\boldsymbol{L}$ and the specifications $T_{5\%}$ and d_{min}*

1. *The damping d_2 is calculated by (4.36).*

2. *According to d_2, the desired open-loop frequency ω_{K} is determined by (4.41), (4.45) or the graphical relation in Fig. 4.6.*

3. *The open-loop gain k is determined by* (4.40).

4. *Use the specified parameters k and ω_{K} for the design of the controller $K(s)$ to shape the open-loop function $G_0(s) = G(s)\,K(s)$ such that it has approximately the form given in* (4.37).

Result: *Local controller $K(s)$.*

The design procedure described in Algorithm 4.1 can equally be performed by the application of the well known root-locus design method. According to Assumption 4.12 and the specifications k and T_{K} of the resulting desired open-loop transfer function (4.37), the root-locus of $G_0(s)$ is required to be composed of two branches belonging to the poles $s_{01} = 0$ and $s_{02} = -\omega_{\mathrm{K}}$. The location of the poles (4.25) of the decoupled error dynamics $\tilde{S}_i(s)$, $(i = 2, N)$, which can be determined by using (4.38) and (4.39), specifies the region of the root locus where the closed-loop poles of $G_0(s)$ in relation to the gain provided by the Laplacian eigenvalues λ_i should lie. Figure 4.4 shows the corresponding region for the location of the poles of the decoupled error dynamics by the shadowed surface, which are related to the network considered in Example 4.1. Hence, in the first step the dynamics of the transfer function $K(s)$ have to be designed such that the root locus of $G_0(s)$ has two branches that are located inside the desired area. In the second step, the control gain has to be chosen to place the closed-loop poles with respect to the Laplacian eigenvalues λ_2 and λ_N appropriately.

4.7 Extensions

4.7.1 Synchronization of agents with non-identical dynamics

The design method presented in the previous section is derived for agents with identical dynamics. In practice the agents are rarely identical, but in most of the cases the agents are similar in the sense that a part of the dynamics is common and the remaining part is individual (cf. Example 4.2).

This section considers the synchronization of non-identical agents

$$Y_i(s) = \underbrace{G(s)\tilde{G}_i(s)}_{G_i(s)} U_i(s) + D_i(s), \qquad i = 1, 2, 3, \ldots, N$$

with the design method presented in the previous section. The decomposition of the multi-agent system, which is addressed in Lemma 4.1, is fundamental for the application of the elaborated design method. Since the decomposition only requires the open-loop transfer functions to be identical, it is straightforward to extend the design approach to non-identical agent dynamics $G_i(s)$ by the introduction of individual local controllers $K_i(s)$, which have to be designed such that the open-loop transfer functions $G_{0i}(s) = G_i(s)K_i(s)$ are identical:

$$G_{0i}(s) = G_0(s), \quad i = 1, 2, \ldots, N.$$

Based on Assumption 4.5 and the corresponding requirement on the open-loop transfer function (4.21) – (4.23), the individual local controllers $K_i(s)$ have to create integral first-order open-loop transfer functions $G_{0i}(s)$ simultaneously. Hence, the design of the controllers $K_i(s) = K(s)\tilde{K}_i(s)$ can be obtained in the following two steps.

(D1) The individual compensating component $\tilde{K}_i(s)$ have to be designed to compensate the individual agent dynamics $\tilde{G}_i(s)$:

$$\tilde{K}_i(s) = \tilde{G}_i^{-1}(s).$$

(D2) The common component $K(s)$ have to be designed with respect to the common agent

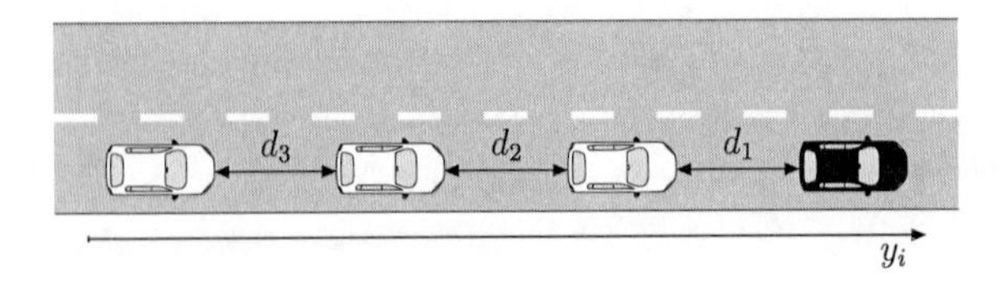

Figure 4.7: Platoon of $N = 4$ vehicles

dynamics $G(s)$ by application of Algorithm 4.1.

Note that only proper transfer functions $\tilde{K}_i(s)$ are suitable for implementation purposes. In the case of non proper transfer functions $\tilde{K}_i(s)$, the required integral-first order behavior have to be approximated.

Example 4.2 *Platooning vehicles*

Consider a platoon of N non-identical vehicles modeled in the frequency domain by

$$G_i(s) = \frac{1}{s\,(m_i\,s + \gamma_i)}, \qquad i = 1,\, 2,\, \ldots,\, N, \tag{4.47}$$

where each vehicle has a different mass m_i and friction coefficient γ_i. The control goal is to ensure constant distances d_i between the vehicles to avoid collisions (Fig. 4.7). This control goal is achieved if the vehicles are synchronized with respect to their velocities and distance shifted positions. Note that even if the vehicles are non-identical synchronization is possible because of some common dynamics, namely, the integrator resulting from the relation between velocity and position. The individual part of the dynamics (4.47) of the i-th agent is represented by the transfer function

$$\tilde{G}_i(s) = \frac{1}{m_i\,s + \gamma_i}.$$

The application of the design steps (D1) – (D2) results in

$$\tilde{K}_i(s) = \tilde{G}_i^{-1}(s) = m_i\,s + \gamma_i \tag{4.48}$$

and

$$K(s) = \frac{k}{T_{\mathrm{K}}\,s + 1}. \tag{4.49}$$

The combination of (4.48) and (4.49) shows that synchronization of non-identical vehicles can be achieved by the use of the individual local controllers, which are given by the proper transfer functions

$$K_i(s) = k\,\frac{m_i\,s + \gamma_i}{T_{\mathrm{K}}\,s + 1}.$$

4.7.2 Synchronization in directed communication networks

The design method presented in Section 4.6 is based on three basic properties of the networked multi-agent system. As mentioned before the decomposition provided in Lemma 4.1 is fundamental for the application of the elaborated design method. Note that the decomposition is only valid for Laplacian matrices that are diagonalizable. Secondly, the time domain behavior of the decoupled synchronization errors $\tilde{S}_i(s)$, $(i = 2, 3, \ldots, N)$ is elaborated under the assumption of real-valued eigenvalues. Thirdly, the synchronization error $\mathbf{e}(t)$ is bounded with respect to Theorem 4.2 if and only if the Laplacian matrix has an orthogonal basis of eigenvectors. This requirement is in general fulfilled for symmetric Laplacian matrices, which is the case for undirected communication networks. However, the design method can also be applied for the synchronization of agents in directed networks if the following assumptions are made.

Assumption 4.7. *The Laplacian matrix $\mathbf{L}$, which is associated with the communication network is assumed to*

1. *have only real valued eigenvalues and*
2. *to be diagonalizable.*

The following theorem is a modification of Theorem 4.2, which provides the bound on the synchronization error $\mathbf{e}(t)$ for directed networks that fulfill Assumption 4.7.

Theorem 4.3 (Bound on the synchronization error in directed networks). *The Euclidean norm of the overall synchronization error $\mathbf{e}(t)$ in a directed network that fulfills Assumption 4.7 is bounded by*

$$\|\mathbf{e}(t)\| \leq \begin{cases} \gamma\, |h_2(t)|\, \|\boldsymbol{d}\|, & d_2 \geq 1 \\ \gamma\, \left|\hat{h}_2(t)\right|\, \|\boldsymbol{d}\|, & d_2 < 1 \end{cases} \tag{4.50}$$

where

$$\gamma = \|\boldsymbol{T}\| \|\boldsymbol{T}^{-1}\|,$$

$\boldsymbol{T}$ the eigenvector matrix of the Laplacian matrix $\mathbf{L}$ and $\mathbf{d} = (d_1, \ldots, d_N)^T$ the magnitude of the step disturbance defined in (4.5).

Proof. The first part of the proof is identical to the proof of Theorem 4.2. From eqn. (4.33) it is easy to see that the synchronization error is bounded by

$$\begin{aligned}\|\boldsymbol{e}(t)\| &\leq \left\|\boldsymbol{T}\,\boldsymbol{H}(t)\,\boldsymbol{T}^{-1}\right\|\,\|\boldsymbol{d}\| \\ &\leq \underbrace{\|\boldsymbol{T}\|\,\|\boldsymbol{T}^{-1}\|}_{\gamma}\,\|\boldsymbol{H}(t)\|\,\|\boldsymbol{d}\|\end{aligned}$$

with $\boldsymbol{H}(t)$ defined in (4.34). The application of the spectral norm of matrices on Lemma 4.3 shows that

$$\begin{aligned}\|\boldsymbol{H}(t)\| &= \sqrt{\max_i\ \lambda_i\left(\boldsymbol{H}^2(t)\right)} \\ &= \sqrt{\max_i\ h_i^2(t)} \\ &= |h_2(t)|\end{aligned}$$

holds. The proof follows directly from Lemma 4.3. □

According to the new result, Corollary 4.2, the eqns. (4.41), (4.45) and the graphical relation in Fig. (4.6) have to be adapted.

Corollary 4.2 (Relation between the bound and the synchronization time for directed networks)**.** *The networked multi-agent system* (4.11) *is synchronized with respect to requirement* (4.2) *if the dynamics of the local controllers $K(s)$ are designed such that Assumption 4.5 is fulfilled and for $t \geq T_{5\%}$ and $d_2 \geq 1$*

$$\gamma\,|h_2(t)| \leq 0.05$$

or in the case that $d_2 < 1$

$$\gamma\,|\hat{h}_2(t)| \leq 0.05$$

holds.

Proof. The proof follows by considering the result of Theorem 4.3 and the proof of Corollary 4.1. □

Analogously to the derivation of the relations (4.41), (4.45) and the graphical relation in Fig. (4.6), the following modified design formulas for the frequency ω_{K} are obtained with respect to Corollary 4.2:

$$\omega_{\mathrm{K}} = \begin{cases} \dfrac{2}{T_{5\%}} \left(3 + \ln(\gamma) - \ln\left(\sqrt{1 - \dfrac{\lambda_N}{\lambda_2} d_{\min}^2} \right) \right), & d_2 < 1 \\ f(T_{5\%},\, d_2,\, \gamma)\,, & 1 \leq d_2 < 2 \\ \dfrac{6 + 2\,\ln(\gamma)}{T_{5\%}} \left(1 - \sqrt{1 - \dfrac{\lambda_2}{\lambda_N\, d_{\min}^2}} \right)^{-1}, & d_2 \geq 2. \end{cases} \tag{4.51}$$

Note that the factor γ has to be considered when numerically determining ω_{K} for the interval $1 \leq d_2 < 2$. These equations turn obviously into the design formulas presented in Section 4.6.4 if $\gamma = 1$. Except for the changed formulas of the frequency ω_{K}, the design method can be applied as it is described in Algorithm 4.1.

4.8 Application example: Synchronization of a vehicle platoon

4.8.1 Vehicle model and control aim

This section demonstrates the behavior of the elaborated design approach by using the experimental plant presented in Section 2.6. For the analysis of the design method, the synchronization of the lateral dynamics (2.25) of four differentially driven robots as shown in Fig. 2.5 is considered.

The networked controller should synchronize the mobile robots from their initial velocities $\dot{y}_i(0) = 0$ and their initial positions $y_i(0) = y_{0i}$, where only the position of the first robot is chosen to be different from zero:

$$\boldsymbol{x}_{01} = \begin{pmatrix} 140 \\ 0 \end{pmatrix}, \quad \boldsymbol{x}_{02} = \begin{pmatrix} 0 \\ 0 \end{pmatrix}, \quad \boldsymbol{x}_{03} = \begin{pmatrix} 0 \\ 0 \end{pmatrix}, \quad \boldsymbol{x}_{04} = \begin{pmatrix} 0 \\ 0 \end{pmatrix}. \tag{4.52}$$

The frequency domain behavior of the model (2.25) with respect to the initial conditions (4.52) is given by

$$Y_i(s) = \frac{1}{22.3\, s\, (1.6\, s + 1)}\, U_i(s) + y_{0i}\, \frac{1}{s}, \quad i = 1,\, 2,\, 3,\, 4.$$

As a special case, initial conditions of the integrator belonging to the model (2.25) can always be interpreted as a step disturbance of the output signal, as illustrated in Fig. 4.2.

4.8.2 Synchronization of vehicles in undirected networks

The interconnection structures that were used in the experiments are shown in Fig. 4.8. This

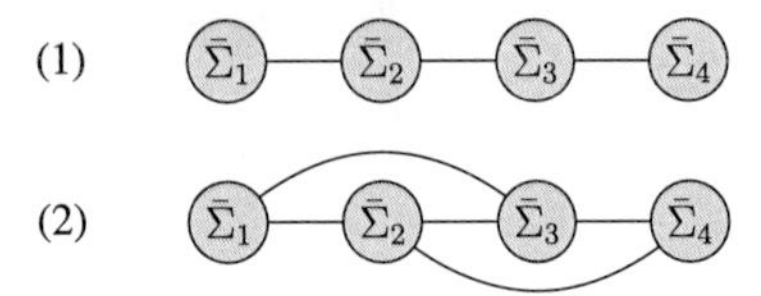

Figure 4.8: Undirected interconnection structure

interconnection structures are chosen to show the influence of additional couplings on the agents behavior. In terms of a vehicle platoon, the first network represents a chain-like structure, where physically neighboring agents are coupled. The second network additionally implements an information exchange between the second-nearest neighbors. The set of nonzero eigenvalues of the corresponding Laplacian matrices

$$\boldsymbol{L}_1 = \begin{pmatrix} 1 & -1 & 0 & 0 \\ -1 & 2 & -1 & 0 \\ 0 & -1 & 2 & -1 \\ 0 & 0 & -1 & 1 \end{pmatrix}, \qquad \boldsymbol{L}_2 = \begin{pmatrix} 2 & -1 & -1 & 0 \\ -1 & 3 & -1 & -1 \\ -1 & -1 & 3 & -1 \\ 0 & -1 & -1 & 2 \end{pmatrix}$$

is given by $\Lambda_1 = \{0.59,\ 2,\ 3.41\}$ and $\Lambda_2 = \{2,\ 4,\ 4\}$, respectively.

Experimental results for network 1

The following table gives specifications for the considered experiments and the controller parameters ω_{K} and k of the controller

$$K(s) = k\,\frac{22.3\,(1.6\,s+1)}{T_{\mathrm{K}}\,s+1},$$

obtained by Algorithm 4.1.

Scenario	$T_{5\%}$ [s]	$d_{\min}$	d_2	ω_{K} [rad/s]	k
1.1	10	0.4	0.97	0.87	0.4
1.2	10	0.9	2.17	5.35	0.48
1.3	8	0.4	0.97	1.09	0.5
1.4	8	0.9	2.17	6.68	0.6

The results are presented in Fig. 4.9 and Fig. 4.10. The experiments indicate that the spec-

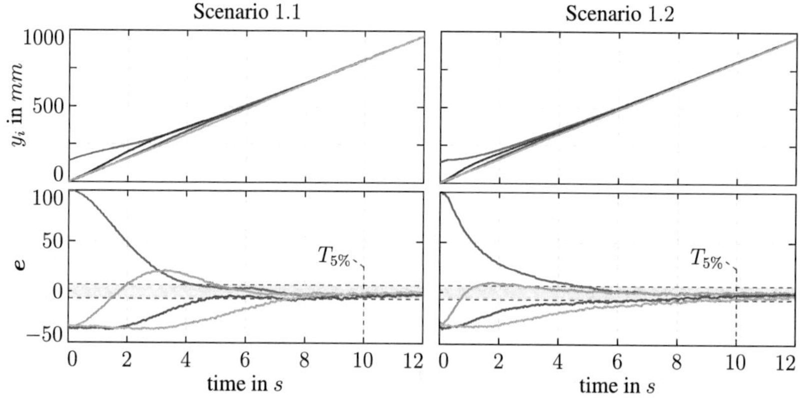

Figure 4.9: Transient behavior of mobile robots coupled by network (1)

ified settling times are satisfied by all elements of the synchronization error $e(t)$. The trajectories of the synchronization error confirm that the damping acts as a suitable tuning factor to adjust the error dynamics. While the errors in Scenario 1.1 and 1.3 exhibit a certain overshoot, a very smooth error trajectory is obtained in Scenario 1.2 and 1.4. The adjustment of the damping has to be accomplished as a trade-off between the error performance and the required control input. However, this is a natural constraint which is well-known from classic control theory.

A comparison of Scenario 1.1 and 1.3 as well as of Scenario 1.2 and 1.4, which exhibit the same damping of the decoupled error dynamics, reveals that different settling times can be interpreted as a scaling of the time axis. If constraints on the control input can be neglected, the qualitative behavior can be set by the damping and the specified settling time can be adjusted separately.

Experimental results for network 2

The following table presents the controller parameters ω_K and k obtained by considering network (2). The corresponding experimental results are presented in Fig. 4.11 and Fig. 4.12. Due to the interconnection structure and the choice of the initial condition, the output signal of the first mobile robot propagates equally onto the second and the third robot. Hence, the robots two and three reveal an identical behavior.

In contrast to network (1), the requirement of the large minimum damping $d_{\min} = 0.9$ in

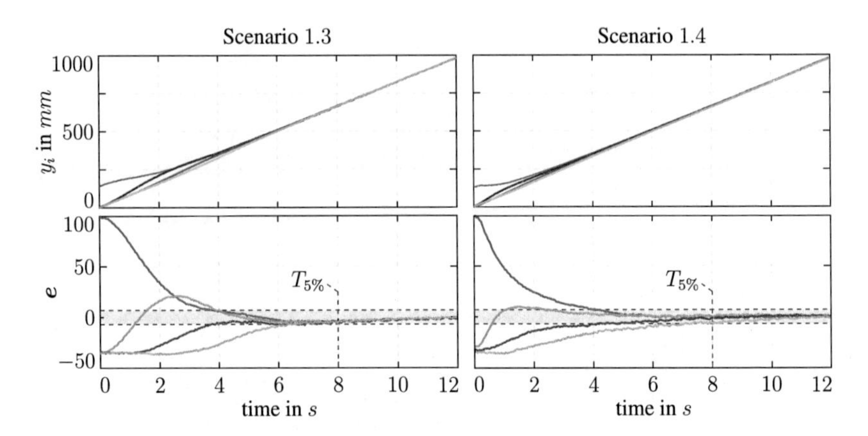

Figure 4.10: Transient behavior of mobile robots coupled by network (1)

Scenario	$T_{5\%}$ $[s]$	d_{min}	d_2	ω_{K} $[\text{rad}/s]$	k
2.1	10	0.4	0.57	0.64	0.25
2.2	10	0.9	1.27	1.75	0.13
2.3	8	0.4	0.57	0.8	0.31
2.4	8	0.9	1.27	2.15	0.17

Scenario 2.2 shows a less significant enlargement of the frequency ω_{K}. This results from the larger ratio $\frac{\lambda_2}{\lambda_N}$. According to the eigenvalues of the Laplacian matrix $\boldsymbol{L}_2$, the poles of the decoupled dynamics are located within a smaller interval. Correspondingly, smaller amplitudes of the control input are observed. From this aspect, network (2) is more favorable than network (1).

4.8.3 Synchronization of vehicles in directed networks

The design of networked controllers for agents coupled by directed communication networks is considered for the leader-flower structure shown in Fig. 4.13. The corresponding Laplacian matrix is given by:

$$\boldsymbol{L}_1 = \begin{pmatrix} 0 & 0 & 0 & 0 \\ -1 & 2 & -1 & 0 \\ 0 & -1 & 1 & 0 \\ 0 & 0 & -1 & 1 \end{pmatrix}.$$

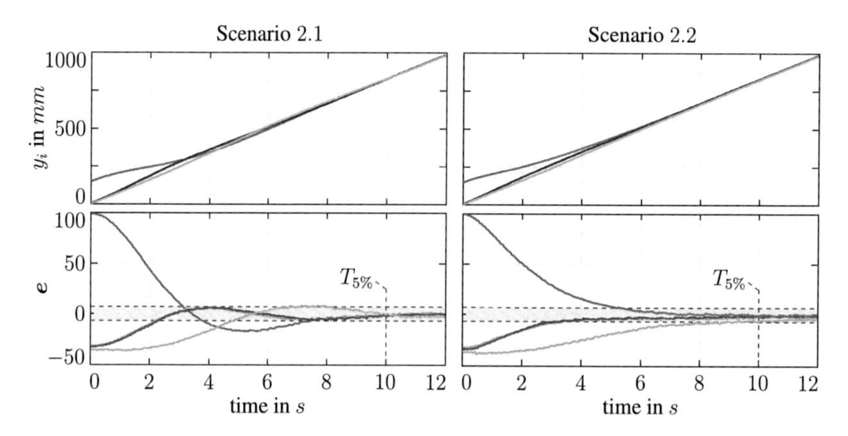

Figure 4.11: Transient behavior of mobile robots coupled by network (2)

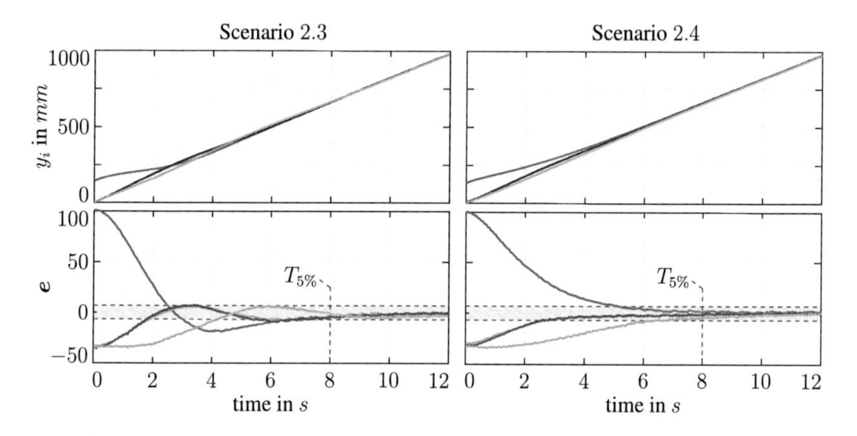

Figure 4.12: Transient behavior of mobile robots coupled by network (2)

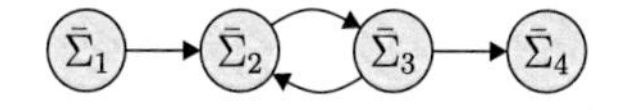

Figure 4.13: Leader-follower communication network

In the considered communication network, the first mobile robot is the leader. The output signal of the leader influences all other robots while it receives no feedback. Consequently, the trajectory of the leader acts as a reference. The set of nonzero eigenvalues of the Laplacian matrix is given by $\Lambda_1 = \{0.38,\ 1,\ 2.62\}$.

For the application of the adapted design method presented in Section 4.7.2, the additional factor γ in Theorem 4.3 has to be determined:

$$\gamma = 6.02.$$

Two different sets of specifications are analyzed for the network. The following table summarizes the corresponding results. The transfer function of the controller has the same structure

Scenario	$T_{5\%}$ [s]	$d_{\min}$	d_2	ω_{K} [rad/s]	k
3.1	16	0.4	1.05	1	0.6
3.2	16	0.8	2.09	4.94	0.74

as in the case of undirected networks. According to the damping $d_2 = 1.05$, the frequency ω_{K} is determined from the diagram in Fig. 4.14, which is specific for $\gamma = 6.02$.

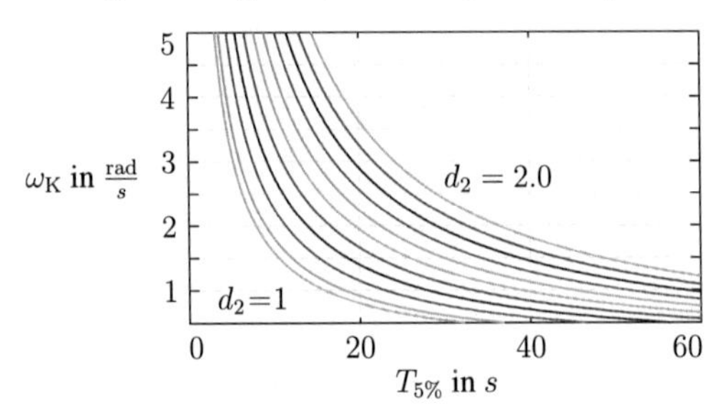

Figure 4.14: Relations between ω_{K}, $T_{5\%}$ and γ for $1 \leq d_2 < 2$

The experimental results in Fig. 4.15 confirm the applicability of the design method to directed networks. In all scenarios, the specified synchronization time is satisfied. As a drawback of the additional factor γ in the bound on the synchronization error, the results show an increased conservatism. This is evident from the fact that the synchronization error decreases below the required bound significantly earlier than the specified synchronization time.

The results also comply with the properties discussed for undirected communication networks. The variation of the specified damping yields the expected effect on the smoothness of the synchronization error as well as on the amplitudes of the control input. Furthermore,

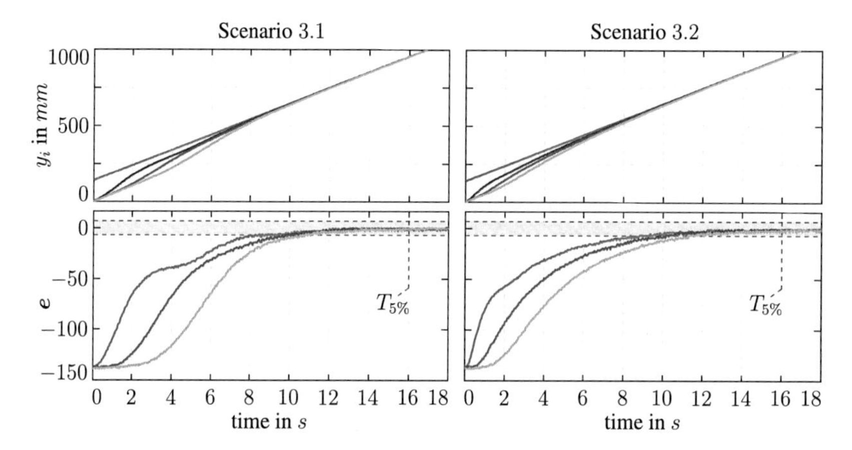

Figure 4.15: Transient behavior of mobile robots in leader-follower configuration

the magnitude of the relation of the Laplacian eigenvalues $\frac{\lambda_2}{\lambda_N}$ shows the same impact on the performance.

Publications

The results concerning the design of dynamic networked controllers for the synchronization through the feedback of output signals were first outlined in [16] and partially presented in [5].

5 Gossip algorithms for the synchronization of multi-agent systems

5.1 Synchronization in point-to-point networks

In the previous two chapters, design methods for synchronizing controllers were presented that can be used for communication networks, where each agent is allowed to exchange signals with all of its neighbors simultaneously. The following investigations are devoted to the development of control algorithms for networks, where each agent can exchange information with only one of its neighbors at the same time (Fig. 5.1). This restriction appears in applications with limited computational, communication and energy resources, where point-to-point or wireless ad-hoc networks are used for information exchange.

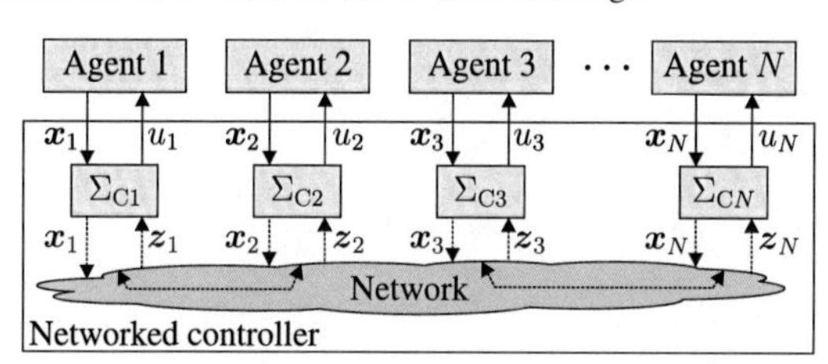

Figure 5.1: Agent network with a pairwise information exchange

For this type of communication networks two control algorithms are presented that ensure asymptotic synchronization of identical discrete-time agents defined by

$$\lim_{k\to\infty} \|\boldsymbol{x}_i(k) - \boldsymbol{x}_j(k)\| = 0, \quad i,\, j, = 1,\, 2,\, \ldots,\, N. \tag{5.1}$$

Figure (5.1) shows the structure of the networked controller, where the decentralized control algorithms Σ_{Ci}, $(i = 1, 2, \ldots, N)$ are only allowed to perform a pairwise information exchange between neighboring agents, which is symbolized by the dashed lines. Both control algorithms work by repeating the following two basic steps:

1. Establish a coupling to a neighbor that is currently not connected to any other agent.

2. Synchronize with the coupled agent in finite time.

The difference between the two control algorithms consists in the choice of the neighboring agents, which is either deterministic or random. A deterministic coupling of the agents is achieved by the use of a periodically repeating coupling sequence. In the case of a random agent coupling, each agent establishes a coupling to one of its neighbors with the same probability. The synchronization of a coupled pair of agents in finite time is achieved by the use of a dead-beat controller. It is shown that a coupled pair of agents reaches synchronization in at most n time steps, where n is the dynamic order of the agents.

The application of the elaborated control algorithms to the synchronization of integrator systems reveals the similarity to gossip algorithms (cf. [102], [103]). However, the results presented in this chapter extend the existing literature by proving necessary and sufficient conditions for the synchronization of discrete-time multi-agent systems with arbitrary linear dynamics.

The analysis of the behavior of the networked agents shows that asymptotic synchronization of unstable agents is achieved if and only if the synchronization of the coupled agents happens faster than the agent's divergence towards their unstable behavior, where the attribute faster is defined precisely later.

For the special case of semi-stable agents, the number $k_{5\%}$ of time steps is estimated, after which the sum of the synchronization errors $\boldsymbol{e}_i(k)$, $(i = 1, 2, \ldots, N-1)$ is bounded by

$$\left(\sum_{i=2}^{N} \| \underbrace{\boldsymbol{x}_1(k) - \boldsymbol{x}_i(k)}_{\boldsymbol{e}_{i-1}(k)} \|^2 \right)^{\frac{1}{2}} \leq 0.05 \, \|\boldsymbol{e}_0\|, \quad \text{for } k \geq k_{5\%}, \tag{5.2}$$

where the initial state of the overall synchronization error is denoted by the stack $\boldsymbol{e}_0^T = (\boldsymbol{e}_1^T(0) \; \boldsymbol{e}_2^T(0) \; \cdots \; \boldsymbol{e}_{N-1}^T(0))$. This estimation shows that after $k_{5\%}$ discrete-time steps practical synchronization with respect to eqn. (5.2) is achieved. The relation between the underlying continuous-time behavior of the agents and the fact that a dead-beat controller is used for the synchronization of the coupled agents in finite time, shows that reducing the sampling time directly reduces the synchronization time of the networked agents. Consequently, a method for the design of the sampling time is developed to guarantee practical synchronization for the given time $T_{5\%}$:

$$\left(\sum_{i=1}^{N-1} \|e_i(t)\|^2 \right)^{\frac{1}{2}} \leq 0.05 \, \|e_0\|, \quad \text{for } t \geq T_{5\%}. \tag{5.3}$$

With this procedure, not only asymptotic synchronization is considered, but also the performance of the transient behavior regarding the requirement on the synchronization time $T_{5\%}$. Furthermore, the relation between the time for synchronization $k_{5\%}$ and the associated number of couplings between the agents is investigated and an answer to the following question is given: How many pairwise agent couplings are sufficient for synchronization?

Structure of this chapter. The agent model and important assumptions are given in Section 5.2. The decentralized control algorithm is derived in Section 5.3, where the synchronization analysis is considered in Section 5.4 for a deterministic agent coupling and in Section 5.5 for a random agent coupling, respectively. Section 5.6 illustrates the control algorithm by its application to the synchronization of harmonic oscillators.

5.2 Agent model and assumptions

The agent dynamics are given by the discrete-time state-space models

$$\Sigma_i : \begin{cases} \boldsymbol{x}_i(k+1) = \boldsymbol{A}\,\boldsymbol{x}_i(k) + \boldsymbol{b}\,u_i(k), & \boldsymbol{x}_i(0) = \boldsymbol{x}_{0i} \\ y_i(k) = \boldsymbol{c}^T \boldsymbol{x}_i(k), & i = 1, 2, \ldots, N \end{cases} \tag{5.4}$$

where

- $\boldsymbol{x}_i(k)$ is the n-dimensional state vector,
- $u_i(k)$ the scalar control input and
- $y_i(k)$ the scalar output signal of the i-th agent.

In general, the synchronization problem is trivially solved for stable agents. This chapter considers non-trivial solutions, where the matrix $\boldsymbol{A}$ is assumed to be unstable.

Assumption 5.1. *The matrix $\boldsymbol{A}$ has at least one eigenvalue outside or on the unit circle of the complex plane:*

$$\rho(\boldsymbol{A}) \geq 1.$$

The unstable behavior of the agents has to be considered explicitly in the design procedure to reach synchronization. The overall aim of the design procedure is the synchronization of the agents towards a synchronous trajectory $y_s(k)$ that corresponds to the averaged behavior of the autonomous agents Σ_i ($u_i(k) = 0$) given by

$$y_s(k) = \boldsymbol{c}^T \boldsymbol{A}^k \boldsymbol{x}_{0s}. \tag{5.5}$$

Thus, the initial value of the synchronous trajectory $\boldsymbol{x}_{0s}$ is the average of the initial states of the agents:

$$\boldsymbol{x}_{0s} = \frac{1}{N} \sum_{i=1}^{N} \boldsymbol{x}_{0i}.$$

As mentioned in Section 5.1 the agents are only allowed to perform a pairwise information exchange. Hence, the following is assumed throughout this chapter.

Assumption 5.2. *Each agent can only exchange information with at most one other agent at the same time. Disjoint pairs of agents can exchange information at the same time.*

Moreover, the agents are assumed to be completely controllable.

Assumption 5.3. *The agents Σ_i are completely controllable*

$$\mathrm{rank} \begin{pmatrix} \boldsymbol{b} & \boldsymbol{A}\boldsymbol{b} & \boldsymbol{A}^2\boldsymbol{b} & \cdots & \boldsymbol{A}^{n-1}\boldsymbol{b} \end{pmatrix} = n.$$

5.3 Synchronizing controller design

5.3.1 Structure of the networked controller

Figure 5.1 illustrates the structure of the networked controller consisting of local control algorithms Σ_{Ci} and the communication network. The coupling partner of the i-th agent can either be chosen deterministically or randomly. A deterministic change of the couplings among the agents leads to a simplified analysis of the agent behavior, which allows the derivation of a necessary and sufficient synchronization condition in Theorem 5.1. In the case of a random agent coupling the behavior of the overall system is analyzed by considering the expected value of the state of the overall system. The resulting synchronization condition refers to the convergence of the expected value of the synchronization errors $\boldsymbol{e}_i(k) = \boldsymbol{x}_1(k) - \boldsymbol{x}_{i+1}(k)$, $(i = 1, \ldots, N-1)$, which is considered in Theorem 5.3.

The design of control algorithms for both types of agent couplings is considered, where the controllers Σ_{Ci} are required

(R1) – to organize the coupling between the pairs of agents and

(R2) – to synchronize a coupled pair of agents in finite time.

The first requirement makes the use of a communication unit (CU) necessary, which creates the couplings among the agents either deterministically or randomly. Figure 5.2 shows the structure of the local control algorithms Σ_{Ci}. The symbol $\bar{\Sigma}_i$ is used for the representation of the controlled i-th agent.

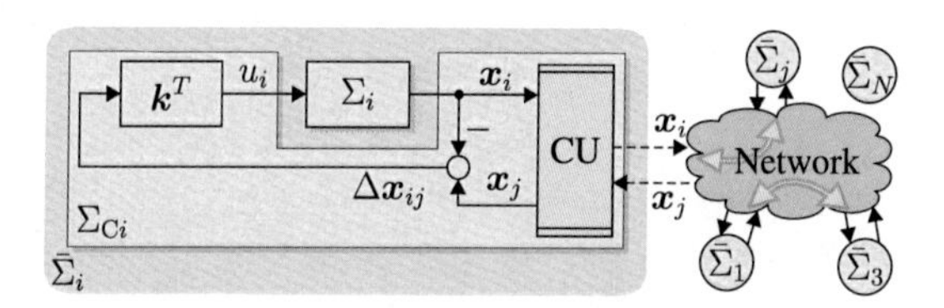

Figure 5.2: Structure of the control algorithms Σ_{Ci}

The idea behind this structure relates on the following two steps. First, the communication unit has to set up a coupling to a neighboring agent that is currently not connected to any of its neighbors. Secondly, a controller k^T has to be designed to achieve synchronization of a coupled pair of agents in finite time.

The coupling between a pair of agents should persist for the time period they need to synchronize. Hence, the communication unit has to check the synchronization process of the coupled pair of agents and to change the coupling partner after the synchronization of the currently coupled pair is completed. A deterministic coupling of the agents is realized with a periodic coupling sequence. In the case of a random agent coupling, each agent chooses randomly one of its neighbors to communicate. For both coupling specifications, a dead-beat controller k^T is used for the synchronization of the coupled agent pair in finite time.

5.3.2 Design of a synchronizing dead-beat controller

This subsection describes the design of a dead-beat controller to achieve synchronization of a coupled pair of agents (Σ_i, Σ_j) in finite time. The synchronizing control law is chosen as the linear state feedback

$$u_i(k) = \begin{cases} \boldsymbol{k}^T(\boldsymbol{x}_{\sigma_i(k)}(k) - \boldsymbol{x}_i(k)) & \text{if} \quad \sigma(k) \neq 0 \\ 0 & \text{else.} \end{cases} \tag{5.6}$$

The signal $\sigma_i(k)$ is used to specify at each time instant k the index of the agent that is currently coupled to agent Σ_i ($\sigma_i(k) : \mathbb{N}^0 \to \mathcal{N}_i \cup \{0\}$). $\mathcal{N}_i$ represents the set of neighbors of agent Σ_i. The value $\sigma_i(k) = 0$ symbolizes that at time k the agent Σ_i is not coupled to any other agent ($u_i(k) = 0$). Synchronization of the pair (Σ_i, Σ_j) is achieved if the controller $\boldsymbol{k}^T$ is designed to stabilize the system matrix $\tilde{\boldsymbol{A}}$ of the synchronization error described by

$$\begin{aligned} \Delta\boldsymbol{x}_{ij}(k+1) &= \boldsymbol{x}_i(k+1) - \boldsymbol{x}_j(k+1) \\ &= \underbrace{(\boldsymbol{A} - 2\,\boldsymbol{b}\,\boldsymbol{k}^T)}_{\tilde{\boldsymbol{A}}}\,\Delta\boldsymbol{x}_{ij}(k). \end{aligned} \tag{5.7}$$

Equation (5.7) is obtained by considering the agent dynamics (5.4) and the control law (5.6) for $\sigma(k) = j$. Synchronization of the error dynamics (5.7) in finite time is achieved if the feedback $\boldsymbol{k}^T$ is designed to make the matrix $\tilde{\boldsymbol{A}}$ nilpotent (cf. [129], [130]):

$$\left(\boldsymbol{A} - 2\,\boldsymbol{b}\,\boldsymbol{k}^T\right)^k = \boldsymbol{O},\ k \geq n. \tag{5.8}$$

Under the assumption of completely controllable agents, there always exists a dead-beat controller

$$\boldsymbol{k}^T = \frac{1}{2}\,\boldsymbol{s}_{\mathrm{c}}^T \boldsymbol{A}^n \tag{5.9}$$

with

$$\boldsymbol{s}_{\mathrm{c}}^T = \begin{pmatrix} 0 & \cdots & 0 & 1 \end{pmatrix} \begin{pmatrix} \boldsymbol{b} & \boldsymbol{A}\boldsymbol{b} & \boldsymbol{A}^2\boldsymbol{b} & \cdots & \boldsymbol{A}^{n-1}\boldsymbol{b} \end{pmatrix}^{-1}$$

that fulfills (5.8).

The application of a dead-beat controller ensures that coupled pairs of agents are synchronized within $k = n$ time steps. Hence the switching signal $\sigma(k)$ takes a constant value in each of the time intervals $k \in [rn, (r+1)n),\ r = 0,\ 1,\ 2,\ \ldots:$

$$\sigma((r+1)n - k) = \sigma(rn), \qquad k = 1,\ 2,\ \ldots,\ n-1. \tag{5.10}$$

The communication unit has therefore to change the couplings between the agents every n

time steps. Thus, the following holds:

$$\boldsymbol{x}_i(rn) - \boldsymbol{x}_{\sigma_i(rn)}(rn) = \mathbf{0}, \quad \sigma_i(rn) \in \mathcal{N}_i, \quad r = 1, 2, 3, \ldots. \tag{5.11}$$

Figure 5.3 illustrates the agent coupling for an example with $n = 2$, where the r-axis is introduced to indicate the time instants at which synchronization of coupled pairs of agents is achieved in finite time.

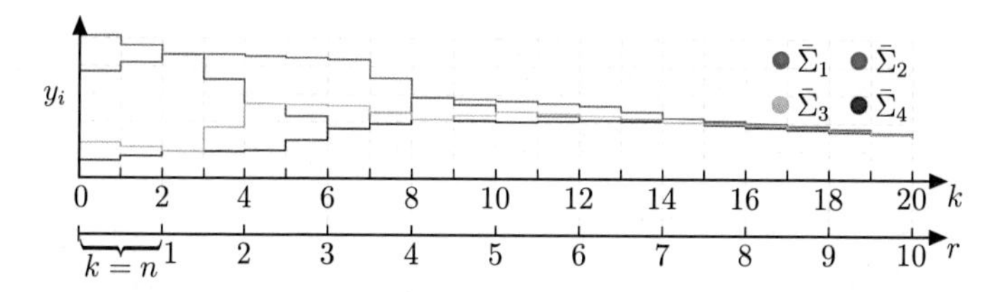

Figure 5.3: Example for the temporal behavior of networked agents with $n = 2$

5.3.3 Behavior of the networked multi-agent system

The behavior of the agents is analyzed by first considering the state-space model of the overall system and its solution, and secondly, by considering the temporal behavior of the overall system at the time instants $k = n, 2n, 3n, \ldots$, described by the r-axis in Fig. 5.3.

The state-space model of the overall multi-agent system is represented in terms of the stack $\boldsymbol{x}^T(k) = \left(\boldsymbol{x}_1^T(k), \ldots, \boldsymbol{x}_N^T(k)\right)$ of the agent states $\boldsymbol{x}_i(k)$ by

$$\boldsymbol{x}(k+1) = (\boldsymbol{I} \otimes \boldsymbol{A})\, \boldsymbol{x}(k) + (\boldsymbol{I} \otimes \boldsymbol{b})\, \boldsymbol{u}(t), \quad \boldsymbol{x}(0) = \boldsymbol{x}_0 \tag{5.12}$$

with $\boldsymbol{u}^T(k) = (u_1(k), \ldots, u_N(k))$. The interconnection of the agents, which is determined by the control law (5.6), is in what follows described by the switching Laplacian matrix $\boldsymbol{L}_{\sigma(k)}$. The control law of the overall system is hence given by

$$\boldsymbol{u}(k) = -(\boldsymbol{L}_{\sigma(k)} \otimes \boldsymbol{k}^T)\, \boldsymbol{x}(k), \quad \sigma(k) \in \mathcal{U}, \tag{5.13}$$

where $\mathcal{U} = \{1, 2, \ldots, U\}$ is set of indices describing the possible interconnection structures. From (5.12) and (5.13), the state equation of the overall closed-loop system is obtained by

$$\boldsymbol{x}(k+1) = (\boldsymbol{I} \otimes \boldsymbol{A} - \boldsymbol{L}_{\sigma(k)} \otimes \boldsymbol{b}\, \boldsymbol{k}^T)\, \boldsymbol{x}(k). \tag{5.14}$$

The representation of the overall system in (5.14) is valid for deterministic and for random couplings. The coupling sequence of the agents is solely a property of the switching signal $\sigma(k)$. For the analysis of the synchronization behavior, it is useful to solve the state equation of the overall closed-loop system (5.14). Particularly, the behavior of the overall closed-loop system (5.14) at the time instants $k = n,\, 2n,\, 3n,\, \ldots$ is significant.

The solution of the overall closed-loop system (5.14) is obtained in three steps. First, a trivial extension of eqn. (5.14) is applied

$$
\begin{aligned}
\boldsymbol{x}(k+1) &= \left(\boldsymbol{I} \otimes \boldsymbol{A} - \boldsymbol{L}_{\sigma(k)} \otimes \boldsymbol{b}\,\boldsymbol{k}^T + \boldsymbol{T}_{\sigma(k)} \otimes \boldsymbol{A} - \boldsymbol{T}_{\sigma(k)} \otimes \boldsymbol{A}\right)\boldsymbol{x}(k) \\
&= \Big[\underbrace{\left(\boldsymbol{I} - \boldsymbol{T}_{\sigma(k)}\right)}_{\frac{1}{2}\boldsymbol{L}_{\sigma(k)}} \otimes \boldsymbol{A} - \boldsymbol{L}_{\sigma(k)} \otimes \boldsymbol{b}\,\boldsymbol{k}^T + \boldsymbol{T}_{\sigma(k)} \otimes \boldsymbol{A}\Big]\,\boldsymbol{x}(k) \\
&= \underbrace{\left[\frac{1}{2}\boldsymbol{L}_{\sigma(k)} \otimes \left(\boldsymbol{A} - 2\boldsymbol{b}\,\boldsymbol{k}^T\right) + \boldsymbol{T}_{\sigma(k)} \otimes \boldsymbol{A}\right]}_{\boldsymbol{M}_{\sigma(k)}}\,\boldsymbol{x}(k),
\end{aligned}
\tag{5.15}
$$

where $\boldsymbol{T}_{\sigma(k)} = \boldsymbol{I} - \frac{1}{2}\boldsymbol{L}_{\sigma(k)}$ is a stochastic matrix of type (2.9). Secondly, the solution of eqn. (5.15) is considered for the time instants $k = rn$, $r = 0,\, 1,\, 2,\, \ldots$, which shows that

$$
\begin{aligned}
\boldsymbol{x}((r+1)n) &= \boldsymbol{M}_{\sigma((r+1)n-1)}\,\boldsymbol{x}((r+1)n-1) \\
&= \boldsymbol{M}_{\sigma((r+1)n-1)}\boldsymbol{M}_{\sigma((r+1)n-2)}\,\boldsymbol{x}((r+1)n-2) \\
&\;\;\vdots \\
&= \boldsymbol{M}_{\sigma((r+1)n-1)}\boldsymbol{M}_{\sigma((r+1)n-2)}\cdots\boldsymbol{M}_{\sigma(rn)}\,\boldsymbol{x}(rn)
\end{aligned}
\tag{5.16}
$$

holds. Remember, a coupling between a pair of agents is established for $k = n$ time steps. For this reason, the coupling structure of the overall system does not change in-between the time instants $k = rn$, $r = 0,\, 1,\, 2,\, \ldots$ and hence

$$
\boldsymbol{x}((r+1)n) = \boldsymbol{M}^n_{\sigma(rn)}\,\boldsymbol{x}(rn) \tag{5.17}
$$

with

$$
\boldsymbol{M}_{\sigma((r+1)n-1)} = \boldsymbol{M}_{\sigma((r+1)n-2)} = \cdots = \boldsymbol{M}_{\sigma(rn)}
$$

holds (cf. eqn. (5.10)). From the second property in eqn. (2.11) and the following computation

$$\begin{aligned}
M^n_{\sigma(rn)} =& \left[\frac{L_{\sigma(rn)}}{2} \otimes \left(A - 2b\,k^T\right) + T_{\sigma(rn)} \otimes A\right]^n \\
=& \left[\frac{L_{\sigma(rn)}}{2} \otimes \left(A - 2b\,k^T\right) + T_{\sigma(rn)} \otimes A\right]\left[\frac{L_{\sigma(rn)}}{2} \otimes \left(A - 2b\,k^T\right) + T_{\sigma(rn)} \otimes A\right]^{n-1} \\
=& \frac{1}{2}\left(I \otimes A - 2b\,k^T\right)\left(L_{\sigma(rn)} \otimes I\right)\left[\frac{L_{\sigma(rn)}}{2} \otimes \left(A - 2b\,k^T\right) + T_{\sigma(rn)} \otimes A\right]^{n-1} + \ldots \\
& \left(I \otimes A\right)\left(T_{\sigma(rn)} \otimes I\right)\left[\frac{L_{\sigma(rn)}}{2} \otimes \left(A - 2b\,k^T\right) + T_{\sigma(rn)} \otimes A\right]^{n-1}
\end{aligned}$$

it is easy to see that

$$\begin{aligned}
M^n_{\sigma(rn)} &= \left[\frac{L_{\sigma(rn)}}{2} \otimes \left(A - 2b\,k^T\right) + T_{\sigma(rn)} \otimes A\right]^n \\
&= \left[\frac{L_{\sigma(rn)}}{2} \otimes \left(A - 2b\,k^T\right)\right]^n + \left(T_{\sigma(rn)} \otimes A\right)^n \\
&= \left[\frac{L^n_{\sigma(rn)}}{2} \otimes \left(A - 2b\,k^T\right)^n\right] + \left(T^n_{\sigma(rn)} \otimes A^n\right).
\end{aligned} \tag{5.18}$$

Equation (5.8) and the first property in eqn. (2.11) allow to reduce eqn. (5.18) to the representation

$$M^n_{\sigma(rn)} = T_{\sigma(rn)} \otimes A^n. \tag{5.19}$$

In the third step, the overall closed-loop system (5.14) is finally rewritten by considering (5.17) and (5.19) to

$$x((r+1)n) = \left(T_{\sigma(rn)} \otimes A^n\right) x(rn), \qquad r = 0,\, 1,\, 2,\, \ldots, \tag{5.20}$$

where the solution of the state equation (5.20) at the time instants $r = 0,\, 1,\, 2,\, \ldots$ is given by

$$x(n) = \left(T_{\sigma(0)} \otimes A^n\right) x_0 \tag{5.21}$$

$$x(2n) = \left(T_{\sigma(n)} \otimes A^n\right) x(n) \tag{5.22}$$

$$\vdots$$

$$x(rn) = \left(T_{\sigma((r-1)n)} \otimes A^n\right) x((r-1)n). \tag{5.23}$$

At this point, the equation of motion of the overall system is obtained from (5.21) – (5.23):

$$\boldsymbol{x}(rn) = \left(\prod_{i=0}^{r-1} \boldsymbol{T}_{\sigma((r-i)n)} \otimes \boldsymbol{A}^{rn} \right) \boldsymbol{x}_0, \quad r = 1,\, 2,\, 3,\, \ldots. \tag{5.24}$$

Equation (5.24) describes the behavior of the overall system without specifying the type of the agent coupling, which is done in the following sections for the deterministic and the random coupling, respectively.

5.4 Deterministic agent coupling

5.4.1 Control algorithm for repetitive synchronization of agent pairs

The requirement (R1) of a deterministic coupling of the agent pairs is achieved by the use of a periodically changing interconnection structure, which is realized with sequences $\mathcal{S}_i$, $(i = 1,\, 2,\, \ldots,\, N)$. A sequence $\mathcal{S}_i$ contains indices of neighboring agents in the order of their interconnection with the i-th agent. Note that the sequences $\mathcal{S}_i$ are referred to as *interconnection sequences* as they provide the order of the temporal coupling between the pairs of coupled agents. The number of different interconnections in a sequence is referred to as the *period* P, $(|\mathcal{S}_i| = P,\; i = 1,\, 2,\, \ldots,\, N)$ and the synchronization process of coupled agents within n time steps as a *synchronization step*. Since only a pairwise coupling of the agents can be realized, it is not always the case that every agent is coupled to one other agent at the same time. A zero entry in the interconnection sequence $\mathcal{S}_i$ of the i-th agent is used if agent Σ_i has no coupling partner during the current synchronization step ($\sigma_i(k) = 0$). The repeating of the couplings between agents according to their interconnection sequences $\mathcal{S}_i$, $(i = 1,\, 2,\, \ldots,\, N)$ satisfies the requirement on a deterministic behavior.

The structure of the *unifying communication graph* $\mathcal{G}$, which results from the union of all couplings between the agents in the P-periodic interconnection sequence, has a large impact on the behavior of the networked agents. In Section 5.4.2 it will be shown that asymptotic synchronization (5.1) of unstable agents Σ_i is possible if and only if the unifying communication graph $\mathcal{G}$ is connected. The focus of this section is on the design of the synchronizing control algorithms $\Sigma_{\mathrm{C}i}$ and the analysis of the overall system behavior, and not on the details of the communication unit design. Therefore, the following assumption is made.

Assumption 5.4. *The interconnection sequences* $\mathcal{S}_i$, $(i = 1, 2, \ldots, N)$ *are given and the corresponding unifying communication graph* $\mathcal{G}$ *is connected.*

In the case of a deterministic agent coupling, the number of Laplacians $\boldsymbol{L}_i$, $(i \in \mathcal{P})$ is determined by the period P, where $\mathcal{P} = \{1, 2, \ldots, P\}$.

Example 5.1 *Periodical agent coupling*

Figure 5.4 shows an example for a periodically repeating coupling sequence of $N = 4$ agents.

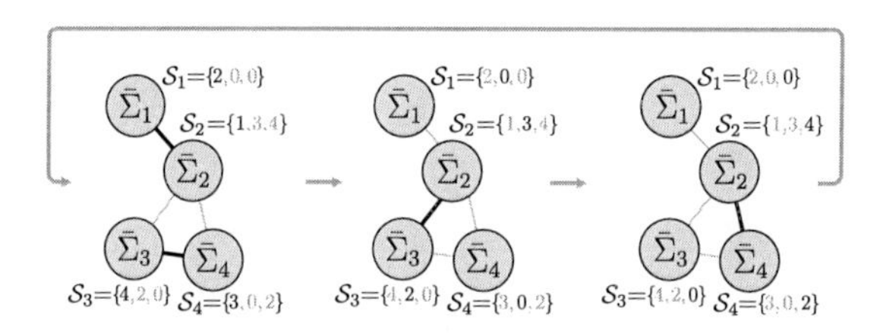

Figure 5.4: Periodically changing interconnection structure

For instance, the interconnection sequence of the third agent, which is given by $\mathcal{S}_3 = \{4, 2, 0\}$ with $P = 3$ shows that its communication unit has to initiate a coupling to agent four and two in the first and the second synchronization step, respectively. The zero entry in $\mathcal{S}_3$ concerns the third synchronization step, in which agent three is not coupled to any other agent. Figure 5.3 shows the corresponding temporal behavior. The κ-axis in Fig. 5.5 is used to indicate the end of a period.

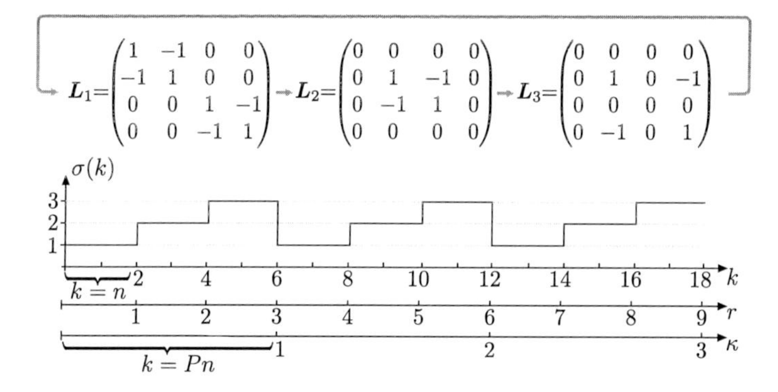

Figure 5.5: Laplacians describing a periodically changing interconnection structure

Figure 5.5 shows the sequence of Laplacians and the corresponding switching signal $\sigma(k)$ of Example 5.1. Note that the switching signal $\sigma(k)$ takes constant values in each of the time intervals $k \in [rn, (r+1)n)$:

$$\sigma(k) = \left\lceil \frac{r+1}{P} \right\rceil, \; r = 0, 1, \ldots .$$

The procedure for establishing a deterministic agent coupling to achieve asymptotic synchronization is summarized in the following algorithm.

Algorithm 5.1. *Control algorithm for a deterministic agent synchronization*

Given: $\mathcal{S}_i$
Initialize: $r = -1$

1. ***Determination of the coupling partner***
   Set $r = r + 1$ *and determine the coupling partner* j *as the* $\lceil (r+1)/P \rceil$*-th element of the coupling sequence* $\mathcal{S}_i$.
   IF $j \neq 0$
   THEN go to step 2
   ELSE set $u_i(k) = 0$, *wait for* n *time steps and go to step 1*
2. ***Connect to a neighboring agent***
   Establish a coupling with neighbor j *and receive* $\boldsymbol{x}_j(k)$.
3. ***Synchronization in finite time***
   Set $u_i(k) = -\boldsymbol{k}^T (\boldsymbol{x}_i(k) - \boldsymbol{x}_j(k))$ *for* n *time steps and go to step 1.*

Result: *Asymptotic agent synchronization through deterministic coupling.*

Algorithm 5.1 ensures synchronization of coupled agents in finite time with the required deterministic coupling sequence. Now, the behavior of the overall system is derived, which is important for the synchronization analysis.

Equation (5.24) describes the behavior of the overall system (5.14) only at the time where the coupled pairs of agents become completely synchronized (cf. eqn. (5.11)). If additionally the periodically changing interconnection structure of the overall system with the P-periodic sequence

$$\boldsymbol{T} = \boldsymbol{T}_P \boldsymbol{T}_{P-1} \cdots \boldsymbol{T}_1$$

is considered, then it follows from (5.24) that

$$\boldsymbol{x}(\kappa P n) = \left(\boldsymbol{T} \otimes \boldsymbol{A}^{Pn}\right)^{\kappa} \boldsymbol{x}_0, \quad \kappa = 0, 1, 2, \ldots \tag{5.25}$$

holds. Equation (5.25) describes, with the new time signal κ, the behavior of the overall system at the time instants $k = 0, Pn, 2Pn, 3Pn, \ldots$, where the P-periodic sequences end

(Fig. 5.5). Since the matrices $\boldsymbol{T}$ and $\boldsymbol{A}^{Pn}$ are constant, it is clear that eqn. (5.25) can be interpreted as the solution of an autonomous discrete-time system. Thus, the system behavior of the overall closed-loop system (5.14) is completely determined by the properties of the matrix $\boldsymbol{T} \otimes \boldsymbol{A}^{Pn}$. Note that $\boldsymbol{T}$ is a doubly stochastic matrix.

5.4.2 Synchronization condition

The main result of this section is a necessary and sufficient condition for asymptotic synchronization of deterministically coupled agents. The result is stated in the following theorem and proved in the appendix.

Theorem 5.1. *The overall system (5.14), which consists of the agents* (5.4) *and the control algorithm 5.1 is asymptotically synchronized if and only if*

$$|\lambda_2|\,\rho\left(\boldsymbol{A}^{Pn}\right) < 1 \tag{5.26}$$

holds. $|\lambda_2|$ *is the second largest magnitude of the eigenvalues of* $\boldsymbol{T}$.

The theorem is interesting because it provides an extension of the current literature on gossip algorithms to the case of the synchronization of general linear discrete-time systems (5.4). In the case of simple integrator systems with $\boldsymbol{A} = 1$, the result of Theorem 5.1 is equal to the result obtained in [103]. In the general case of semi-stable systems ($\rho(\boldsymbol{A}^{Pn}) \leq 1$) synchronization can only be achieved if all eigenvalues apart from the eigenvalue $\lambda_1 = 1$ of the matrix $\boldsymbol{T}$ have magnitudes strictly less than one. This is true if the union of the periodically changing interconnection structures, which is described by the Laplacian matrix $\boldsymbol{L} = \boldsymbol{L}_1 + \boldsymbol{L}_2 + \ldots + \boldsymbol{L}_P$, represents a connected network (cf. [103]). Asymptotically stable systems with $\rho(\boldsymbol{A}^{Pn}) < 1$ are synchronized because $|\lambda_2| \leq 1$ is generally satisfied.

The synchronization condition (5.26) can also be applied to the more interesting case of unstable agents. Since then $\rho(\boldsymbol{A}^{Pn}) > 1$ holds, Theorem 5.1 shows that $|\lambda_2|$ has to be smaller than $1/\rho(\boldsymbol{A}^{Pn})$ to guarantee asymptotic synchronization. As the magnitude $|\lambda_2|$ is a consequence of the network structure and the choice of the agent couplings described by the interconnection sequences $\mathcal{S}_i$, ($i = 1, 2, \ldots, N$), the changing of the interconnection order of the agents directly influences $|\lambda_2|$.

If the interconnection sequences are fixed, $|\lambda_2|$ is fixed as well. Then it is also possible to reduce the sampling time T of the discrete-time agents Σ_i to satisfy eqn. (5.26) and to reach asymptotic synchronization. This is evident by considering the relation between the eigenvalues λ_{ci} of the corresponding continuous-time system and the eigenvalues $\lambda_i(\boldsymbol{A})$ of the discretized system: $\lambda_i(\boldsymbol{A}) = \mathrm{e}^{\lambda_{ci}T}$. Thus, reducing the sampling time T reduces the spectral radius $\rho(\boldsymbol{A}^{Pn})$.

Remark 5.1. *The choice of the sampling time T has an interesting interpretation in the context to the behavior of the overall system. Since a dead-beat controller is used for the synchronization of coupled agent pairs, a reduction of the sampling time T directly reduces the time after which the coupled agents are synchronized: $k = n$ and $k = 0, T, 2T, \ldots$. Hence, synchronization of unstable agents with the control structure described in Section 5.3.1 can only be achieved if the synchronization between the coupled agent pairs is faster than the divergence of the unstable agents.*

The discussion above clearly shows that the convergence rate of the overall closed-loop system (5.14) towards the synchronous trajectory $y_{\mathrm{s}}(k)$ is completely determined by the properties of the matrices $\boldsymbol{T}$ and $\boldsymbol{A}^{Pn}$. Furthermore, it is clear from (2.12), (5.25) and the following computation

$$\begin{aligned}
\lim_{\kappa\to\infty} \boldsymbol{x}(\kappa Pn) &= \left(\lim_{\kappa\to\infty} \boldsymbol{T}^{\kappa} \otimes \lim_{\kappa\to\infty} \boldsymbol{A}^{\kappa Pn}\right)\boldsymbol{x}_0 \\
&= \left(\frac{1}{N}\mathbb{1}\mathbb{1}^T \otimes \lim_{\kappa\to\infty} \boldsymbol{A}^{\kappa Pn}\right)\boldsymbol{x}_0 \\
&= \left(\mathbb{1} \otimes \lim_{\kappa\to\infty} \boldsymbol{A}^{\kappa Pn}\frac{1}{N}\sum_{i=1}^{N}\boldsymbol{x}_{0i}\right)
\end{aligned}$$

that the synchronous trajectory $y_{\mathrm{s}}(k)$ is given by eqn. (5.5).

5.4.3 Design of the sampling time to guarantee performance

The previous section focused on a necessary and sufficient condition for asymptotic synchronization of the networked multi-agent system, consisting of the agent dynamics Σ_i and the controller given in Algorithm 5.1. Furthermore, it is shown that reducing the sampling time T directly reduces the time for the synchronization of a coupled pair of agents. The question of interest in this section is: How to choose the sampling time T in order to satisfy the requirement on the synchronization time (5.3) for any given time $T_{5\%}$. Note that this question is posed only for semi-stable agents.

Assumption 5.5. *The agents Σ_i are assumed to be semi-stable ($\rho(\boldsymbol{A}) = 1$).*

The answer to this question is obtained in two steps. First, the estimation of the number of time steps $k_{5\%}$ is made, after which practical synchronization with respect to eqn. (5.2) is achieved. This result is stated in Lemma 5.1. In the second step, the estimation is used for the proof of Theorem 5.2 in which the sampling time T is determined to satisfy (5.3).

The overall representation of the sum of the synchronization errors in (5.2) is obtained by

$$\begin{aligned}\Bigg(\sum_{i=2}^{N}\|\underbrace{\boldsymbol{x}_1(k)-\boldsymbol{x}_i(k)}_{\boldsymbol{e}_{i-1}(k)}\|^2\Bigg)^{\frac{1}{2}} &= \left\|\begin{matrix}\boldsymbol{x}_1(k)-\boldsymbol{x}_2(k)\\ \vdots \\ \boldsymbol{x}_1(k)-\boldsymbol{x}_N(k)\end{matrix}\right\| \\ &= \|\underbrace{(\boldsymbol{N}\otimes\boldsymbol{I})\,\boldsymbol{x}(k)}_{\boldsymbol{e}(k)}\|,\end{aligned} \tag{5.27}$$

where

$$\boldsymbol{N} = \begin{pmatrix}1 & -1 & 0 & \cdots & 0\\ 1 & 0 & -1 & \ddots & \vdots\\ \vdots & \vdots & \ddots & \ddots & 0\\ 1 & 0 & \cdots & 0 & -1\end{pmatrix}.$$

The corresponding behavior of $\boldsymbol{e}(k)$ at the time instants $k = 0,\, Pn,\, 2Pn,\, 3Pn,\, \ldots$ is obtained by considering eqn. (5.25):

$$\boldsymbol{e}(\kappa Pn) = (\boldsymbol{N}\otimes\boldsymbol{I})\left(\boldsymbol{T}\otimes\boldsymbol{A}^{Pn}\right)^{\kappa}\boldsymbol{x}_0, \quad \kappa = 0,\,1,\,2,\,\ldots\,. \tag{5.28}$$

Since $\boldsymbol{T}$ is a doubly stochastic matrix, it is easy to show that the similarity transformation with the involutory matrix

$$\boldsymbol{V} = \begin{pmatrix}1 & 0 & 0 & \cdots & 0\\ 1 & -1 & 0 & \cdots & 0\\ 1 & 0 & -1 & \ddots & \vdots\\ \vdots & \vdots & \ddots & \ddots & 0\\ 1 & 0 & \cdots & 0 & -1\end{pmatrix}, \quad \boldsymbol{V}^2 = \boldsymbol{I}, \quad \boldsymbol{N} = (\,\boldsymbol{0}\,|\,\boldsymbol{I}\,)\,\boldsymbol{V} \tag{5.29}$$

leads to a decomposition of the doubly stochastic matrix $\boldsymbol{T}$ into an upper block triangular matrix given by

$$\boldsymbol{V}\boldsymbol{T}\boldsymbol{V} = \left(\begin{array}{c|ccc}1 & * & \cdots & *\\ \hline 0 & & & \\ \vdots & & \tilde{\boldsymbol{T}} & \\ 0 & & & \end{array}\right). \tag{5.30}$$

In the special case of synchronization of semi-stable agents, it follows directly from the synchronization condition in Theorem 5.1 that

$$\rho\left(\tilde{\boldsymbol{T}}\right) = |\lambda_2| < 1$$

holds.

The application of the transformation (5.30) to eqn. (5.28) shows that the behavior of the synchronization error $\boldsymbol{e}(k)$ can be rewritten as

$$\begin{aligned}
\boldsymbol{e}(\kappa Pn) &= (\boldsymbol{N} \otimes \boldsymbol{I})\left(\boldsymbol{T}^{\kappa} \otimes \boldsymbol{A}^{\kappa Pn}\right)\boldsymbol{x}_0 \\
&= (\boldsymbol{N} \otimes \boldsymbol{I})\left(\boldsymbol{V}^2\,\boldsymbol{T}^{\kappa}\,\boldsymbol{V}^2 \otimes \boldsymbol{A}^{\kappa Pn}\right)\boldsymbol{x}_0 \\
&= (\boldsymbol{N}\,\boldsymbol{V} \otimes \boldsymbol{I})\left(\boldsymbol{V}\,\boldsymbol{T}\,\boldsymbol{V} \otimes \boldsymbol{A}^{Pn}\right)^{\kappa}(\boldsymbol{V} \otimes \boldsymbol{I})\,\boldsymbol{x}_0 \\
&= ((\boldsymbol{0}\,|\,\boldsymbol{I}\,) \otimes \boldsymbol{I})\left(\left(\begin{array}{c|ccc} 1 & * & \cdots & * \\ \hline 0 & & & \\ \vdots & & \tilde{\boldsymbol{T}}^{\kappa} & \\ 0 & & & \end{array}\right) \otimes \boldsymbol{A}^{\kappa Pn}\right)(\boldsymbol{V} \otimes \boldsymbol{I})\,\boldsymbol{x}_0 \\
&= \left(\tilde{\boldsymbol{T}}^{\kappa} \otimes \boldsymbol{A}^{\kappa Pn}\right)((\boldsymbol{0}\,|\,\boldsymbol{I}\,)\,\boldsymbol{V} \otimes \boldsymbol{I})\,\boldsymbol{x}_0 \\
&= \left(\tilde{\boldsymbol{T}}^{\kappa} \otimes \boldsymbol{A}^{\kappa Pn}\right)\underbrace{(\boldsymbol{N} \otimes \boldsymbol{I})\,\boldsymbol{x}_0}_{\boldsymbol{e}(0)\,=\,\boldsymbol{e}_0}, \qquad \kappa = 0,\,1,\,2,\,\ldots\,.
\end{aligned} \tag{5.31}$$

Note, the relation between the system matrix of the discrete-time agents Σ_i and the underlying continuous-time matrix $\boldsymbol{A}_\mathrm{c}$ is described by $\boldsymbol{A} = \mathrm{e}^{\boldsymbol{A}_\mathrm{c}T}$, where T is the sampling time. The estimation of the synchronization time $k_{5\%}$, is obtained by considering (5.2), (5.27) and (5.31). The substitution of (5.27) and (5.31) into (5.2) gives the inequality

$$\|\boldsymbol{e}(\kappa Pn)\| = \left\|\left(\tilde{\boldsymbol{T}}^{\kappa} \otimes \boldsymbol{A}^{\kappa Pn}\right)\boldsymbol{e}_0\right\| \leq 0.05\|\boldsymbol{e}_0\|, \tag{5.32}$$

which is satisfied with

$$\boldsymbol{A}^{\kappa Pn} = \underbrace{\left(\mathrm{e}^{\boldsymbol{A}_\mathrm{c}n}\right)}_{\tilde{\boldsymbol{A}}}{}^{PT\kappa}$$

if

$$\begin{aligned}
\left\|\left(\tilde{\boldsymbol{T}}^{\kappa} \otimes \boldsymbol{A}^{\kappa Pn}\right)\boldsymbol{e}_0\right\| &\leq \left\|\tilde{\boldsymbol{T}}^{\kappa}\right\|\left\|\tilde{\boldsymbol{A}}^{PT\kappa}\right\|\|\boldsymbol{e}_0\| \\
\left\|\tilde{\boldsymbol{T}}^{\kappa}\right\|\left\|\tilde{\boldsymbol{A}}^{PT\kappa}\right\| &\leq 0.05
\end{aligned} \tag{5.33}$$

holds. An estimation of the matrix norms in (5.33) with respect to the similarity transformations $\boldsymbol{V}_{\mathrm{b}}^{-1}\,\tilde{\boldsymbol{T}}\,\boldsymbol{V}_{\mathrm{b}} = \operatorname{diag}(\lambda_i(\tilde{\boldsymbol{T}}))$ and $\boldsymbol{V}_{\mathrm{a}}^{-1}\,\tilde{\boldsymbol{A}}\,\boldsymbol{V}_{\mathrm{a}} = \operatorname{diag}(\lambda_i(\tilde{\boldsymbol{A}}))$ shows that

$$\left\|\tilde{\boldsymbol{T}}^{\kappa}\right\| \leq \|\boldsymbol{V}_{\mathrm{b}}\|\,\left\|\boldsymbol{V}_{\mathrm{b}}^{-1}\right\|\underbrace{\rho\left(\tilde{\boldsymbol{T}}\right)^{\kappa}}_{=|\lambda_2|^{\kappa}} \tag{5.34}$$

and

$$\left\|\tilde{\boldsymbol{A}}^{PT\kappa}\right\| \leq \|\boldsymbol{V}_{\mathrm{a}}\|\,\left\|\boldsymbol{V}_{\mathrm{a}}^{-1}\right\|\underbrace{\rho\left(\tilde{\boldsymbol{A}}\right)^{PT\kappa}}_{=1} \tag{5.35}$$

holds. Assumption 5.5 leads to the relation $\rho\left(\tilde{\boldsymbol{A}}\right) = 1$, which is essential for the estimation in eqn. (5.35).

Remark 5.2. *From the combination of eqns.* (5.32)–(5.35) *which is given by*

$$\frac{\|\mathbf{e}(\kappa P n)\|}{\|\boldsymbol{e}_0\|} \leq \|\boldsymbol{V}_{\mathrm{a}}\|\,\left\|\boldsymbol{V}_{\mathrm{a}}^{-1}\right\|\,\|\boldsymbol{V}_{\mathrm{b}}\|\,\left\|\boldsymbol{V}_{\mathrm{b}}^{-1}\right\|\,|\lambda_2|^{\kappa}$$

it is easy to see that the convergence rate of the synchronization error $\mathbf{e}(k)$ is solely a property of the interconnection sequence of the agents, which corresponds to $|\lambda_2|$.

Equation (5.33), (5.34) and (5.35) complete the preliminaries for the result of the following lemma.

Lemma 5.1. *A network of semi-stable agents that fulfills the synchronization condition* (5.26) *is synchronized with respect to inequality* (5.2) *after $k_{5\%} = P\,n\,\kappa_{5\%}$ time steps, where*

$$\kappa_{5\%} = \left\lceil \log^{-1}\left(|\lambda_2|\right)\log\left(\frac{0.05}{\|\boldsymbol{V}_{\mathrm{a}}\|\,\left\|\boldsymbol{V}_{\mathrm{a}}^{-1}\right\|\,\|\boldsymbol{V}_{\mathrm{b}}\|\,\left\|\boldsymbol{V}_{\mathrm{b}}^{-1}\right\|}\right)\right\rceil . \tag{5.36}$$

Proof. Inequality (5.2) is equivalent to the representation in eqn. (5.32), which is fulfilled if (5.33) holds. The combination of eqn. (5.33) and the estimations (5.34) and (5.35) shows that (5.2) is fulfilled if

$$\|\boldsymbol{V}_{\mathrm{a}}\|\,\left\|\boldsymbol{V}_{\mathrm{a}}^{-1}\right\|\,\|\boldsymbol{V}_{\mathrm{b}}\|\,\left\|\boldsymbol{V}_{\mathrm{b}}^{-1}\right\|\,|\lambda_2|^{\kappa_{5\%}} \leq 0.05$$

or the equivalent relation

$$\kappa_{5\%} \log\left(|\lambda_2|\right) \leq \log\left(\frac{0.05}{\|\boldsymbol{V}_{\mathrm{a}}\| \, \|\boldsymbol{V}_{\mathrm{a}}^{-1}\| \, \|\boldsymbol{V}_{\mathrm{b}}\| \, \|\boldsymbol{V}_{\mathrm{b}}^{-1}\|}\right).$$

With respect to the synchronization condition (5.26) and the property $\rho(\boldsymbol{A}) = 1$ of a semi-stable system matrix $\boldsymbol{A}$, it is easy to show that $\log\left(|\lambda_2|\right) < 0$ and

$$\kappa_{5\%} \geq \log^{-1}\left(|\lambda_2|\right) \log\left(\frac{0.05}{\|\boldsymbol{V}_{\mathrm{a}}\| \, \|\boldsymbol{V}_{\mathrm{a}}^{-1}\| \, \|\boldsymbol{V}_{\mathrm{b}}\| \, \|\boldsymbol{V}_{\mathrm{b}}^{-1}\|}\right) \tag{5.37}$$

holds. The right side of inequality (5.37) proofs the lemma. The ceiling function in (5.36) is used to create an integer-valued upper bound. □

Remark 5.3. *The result of Lemma 5.1 concerns the analysis of the rate of convergence of deterministically coupled agents, where the number of time steps $k_{5\%}$ is estimated, after which synchronization with respect to inequality* (5.2) *is achieved. Since a dead-beat controller is used and synchronization of coupled agents is achieved in $k = n$ time steps, there exists a direct connection between the time for synchronization $k_{5\%}$ and the associated number of couplings between the agents. According to eqn.* (5.36)*, the number of couplings for synchronization with respect to inequality* (5.2) *is determined by*

$$N_{5\%} = \kappa_{5\%} \, N_{\kappa} \tag{5.38}$$

where N_{κ} is the number of couplings in a P-periodic communication sequence (e.g. $N_{\kappa} = 4$ in Example 5.1). Equation (5.38) *gives an answer to the question: How many couplings are sufficient for synchronization of periodically coupled agents?*

Furthermore, the estimation in eqn. (5.36) is elaborated to be independent of the sampling time T. In particular, this is true for semi-stable agents, which can be seen from (5.33), (5.34), (5.35) and the fact that the matrix $\tilde{\boldsymbol{T}}$ is only characterized by the topology of the communication network and the coupling sequence of the agents. The estimation in eqn. (5.36) solely depends upon the properties of the communication network given by $\tilde{\boldsymbol{T}}$, $(|\lambda_2|, \; \boldsymbol{V}_{\mathrm{b}})$ and the agent dynamics expressed by the eigenvector matrix $\boldsymbol{V}_{\mathrm{a}}$ of $\tilde{\boldsymbol{A}} = \mathrm{e}^{\boldsymbol{A}_{\mathrm{c}} n}$. Hence, $k_{5\%} = P \, n \, \kappa_{5\%}$ is independent of the sampling time T too. This property is important for the determination of the sampling time T in order to satisfy the requirement on the synchronization time given in (5.3).

Equation $t = k T$ describes the relation between the sampling time T, the discrete-time steps k and the corresponding continuous-time t. The estimation in Lemma 5.1 shows that synchronization with respect to (5.3) is achieved if

$$k_{5\%} T \leq T_{5\%} \tag{5.39}$$

for the given synchronization time $T_{5\%}$. Since the estimated number of the synchronization steps $k_{5\%}$ is independent of the sampling time T, eqn. (5.39) can directly be used for the determination of the sampling time T to satisfy the requirement (5.3) for any given synchronization time $T_{5\%}$. This result is summarized in the following theorem.

Theorem 5.2. *Semi-stable agents that fulfill the synchronization condition* (5.26) *are synchronized with respect to inequality* (5.3) *for the given synchronization time* $T_{5\%}$ *if*

$$T \leq \frac{T_{5\%}}{k_{5\%}}.$$

$k_{5\%}$ *is the estimation given in Lemma 5.1.*

5.5 Random agent coupling

5.5.1 Control algorithm for random synchronization of agent pairs

The randomized coupling of the agents is achieved if the communication unit is designed to create the couplings to neighboring agents with probability p_{ij}. In the first step, it is assumed that each agent can be activated with probability $1/N$. In the second step, the activated agent chooses one of its *available* neighbors with probability $p_{ij} = 1/N_i$ to communicate, where $N_i = |\mathcal{N}_i|$ denotes the number of *available* neighbors of agent Σ_i. If an agent is not coupled to another agent, then this agent is said to be *available*. The process of the i-th agent to decide with whom to communicate is assumed to be instantaneous. This means that each agent connects to one of its uncoupled neighbors until the coupling is created, or until all of its neighbors are coupled to another agent within a negligible time. In summary, the following assumptions are made.

Assumption 5.6.

1. *Each agent is activated with probability* $1/N$.

2. *If agent* i *is activated, then the pair* $(\bar{\Sigma}_i, \bar{\Sigma}_j)$ *is coupled with probability* p_{ij}.

3. *The coupling process of the agents is assumed to be instantaneous.*

Example 5.2 illustrates the corresponding coupling behavior.

Example 5.2 *Random agent coupling*

Figure 5.6 shows an example for the random coupling of $N = 4$ agents. The left part of

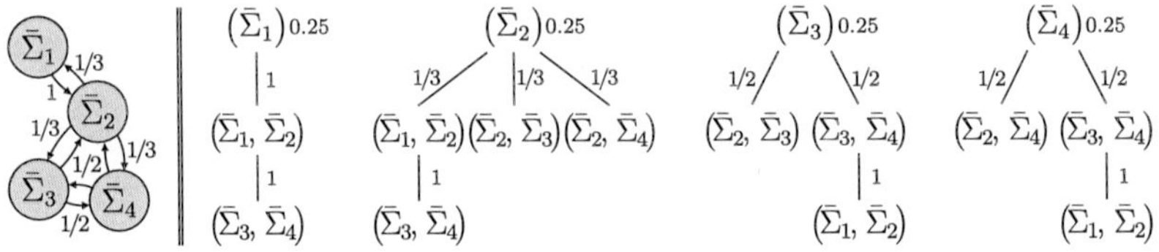

Figure 5.6: Example: Interconnection structures of randomly coupled agents

the figure shows a directed graph, where an edge from $\bar{\Sigma}_i$ to $\bar{\Sigma}_j$ is labeled with the probability belonging to the coupling of agent Σ_i to agent Σ_j. The right side of Fig. 5.6 shows the possible combinations of the couplings between the agents. In the situation where agent $\bar{\Sigma}_2$ is activated as first, a coupling to $\bar{\Sigma}_1$, $\bar{\Sigma}_3$ or $\bar{\Sigma}_4$ can be created respectively with probability $1/3$. If the coupling between the pair $(\bar{\Sigma}_2, \bar{\Sigma}_3)$ or the pair $(\bar{\Sigma}_2, \bar{\Sigma}_4)$ is created, then there do not exist any other agents that can be coupled. If in contrast the pair $(\bar{\Sigma}_2, \bar{\Sigma}_1)$ is coupled, then $\bar{\Sigma}_3$ and $\bar{\Sigma}_4$ can afterwards be coupled with probability 1. The resulting three possible interconnection structures are described by the following Laplacian matrices:

$$\boldsymbol{L}_1 = \begin{pmatrix} 1 & -1 & 0 & 0 \\ -1 & 1 & 0 & 0 \\ 0 & 0 & -1 & 1 \\ 0 & 0 & 1 & -1 \end{pmatrix}, \quad \boldsymbol{L}_2 = \begin{pmatrix} 0 & 0 & 0 & 0 \\ 0 & 1 & -1 & 0 \\ 0 & -1 & 1 & 0 \\ 0 & 0 & 0 & 0 \end{pmatrix}, \quad \boldsymbol{L}_3 = \begin{pmatrix} 0 & 0 & 0 & 0 \\ 0 & 1 & 0 & -1 \\ 0 & 0 & 0 & 0 \\ 0 & -1 & 0 & 1 \end{pmatrix}. \tag{5.40}$$

The probability p_i for each interconnection structure (5.40) that can be attained, is calculated by considering the right side of Fig. 5.6:

$$p_1 = \frac{14}{24}, \quad p_2 = \frac{5}{24}, \quad p_3 = \frac{5}{24}. \tag{5.41}$$

The following algorithm summarizes the procedure for establishing random agent couplings, which are used to achieve synchronization of coupled agents in finite time.

Algorithm 5.2. *Random pairwise agent synchronization*

Given: $\mathcal{N}_i$

Initialize: $\bar{\mathcal{N}}_i = \mathcal{N}_i$

1. ***Determination of the coupling partner***
   Choose a neighbor j from the set $\bar{\mathcal{N}}_i$ with probability $p_{ij} = 1/|\bar{\mathcal{N}}_i|$.

2. ***Connect to a neighboring agent***
   Establish a coupling with neighbor j and receive $\boldsymbol{x}_j(k)$.
   IF *coupling is established*
   THEN *go to step 3*
   ELSEIF $|\bar{\mathcal{N}}_i| = 1$
   THEN *set $\bar{\mathcal{N}}_i = \mathcal{N}_i$, wait for n time steps and go to step 1*
   ELSE *set $\bar{\mathcal{N}}_i = \bar{\mathcal{N}}_i \backslash \{j\}$ and go to step 1*

3. ***Synchronization in finite time***
   Set $u_i(k) = -\boldsymbol{k}^T(\boldsymbol{x}_i(k) - \boldsymbol{x}_j(k))$ for n time steps and go to step 1.

Result: *Asymptotic agent synchronization through random coupling.*

For the representation of the overall system, each interconnection structure is described by a Laplacian matrix $\boldsymbol{L}_i \in \mathcal{R} = \{1, 2, \ldots, R\}$. The number of all possible interconnection structures is denoted by R. Furthermore, each interconnection structure $\boldsymbol{L}_i$ is assigned with probability p_i, where $\sum_{i=1}^{R} p_i = 1$ holds. Consequently, the behavior of the overall system can be expressed by eqn. (5.14) and eqn. (5.24), where the value of the switching signal $\sigma(k) \in \mathcal{R}$ is assigned with the corresponding probability p_i. The behavior of the overall system is difficult to analyze because of the non-deterministic agent coupling. In [102] the expected value of the randomized coupled systems was used for the analysis of the *convergence in expectation*. The same approach is applied in this section. The expected behavior of the overall system (5.24) is represented by

$$\begin{aligned}\mathrm{E}(\boldsymbol{x}(rn)) &= \left(\prod_{i=0}^{r-1} \mathrm{E}\left(\boldsymbol{T}_{\sigma((r-i)n)}\right) \otimes \boldsymbol{A}^{rn}\right) \boldsymbol{x}_0, \\ &= \left(\prod_{i=0}^{r-1} \bar{\boldsymbol{T}} \otimes \boldsymbol{A}^{rn}\right) \boldsymbol{x}_0, \\ &= \left(\bar{\boldsymbol{T}} \otimes \boldsymbol{A}^{n}\right)^r \boldsymbol{x}_0, \quad r = 0, 1, 2, \ldots, \end{aligned} \tag{5.42}$$

where

$$\bar{\boldsymbol{T}} = \mathrm{E}\left(\boldsymbol{T}_{\sigma(k)}\right) = \boldsymbol{I} - \frac{1}{2}\mathrm{E}\left(\boldsymbol{L}_{\sigma(k)}\right) = \boldsymbol{I} - \frac{1}{2}\sum_{i=1}^{R} p_i\,\boldsymbol{L}_i \tag{5.43}$$

is the average of the stochastically independent matrices $\boldsymbol{T}_{\sigma(k)} = \boldsymbol{I} - \frac{1}{2}\boldsymbol{L}_{\sigma(k)}$. Note that $\bar{\boldsymbol{T}}$ is a symmetric doubly stochastic matrix. Equation (5.42) describes the expected behavior of the overall system at the time instants $k = n,\, 2n,\, 3n, \ldots$, at which the dead-beat controller (5.6) satisfies complete synchronization of coupled pairs of agents. Furthermore, (5.42) corresponds to the solution of an autonomous discrete-time system, which is completely determined by the initial state $\boldsymbol{x}_0$ and the properties of the matrix $\bar{\boldsymbol{T}} \otimes \boldsymbol{A}^n$. This result is similar to the case of a deterministic agent coupling. However, for randomly coupled agents the synchronization condition cannot be derived from considering the requirement (5.1). Instead, synchronization of the expected agent behavior as given in the following definition is considered.

Definition 5.1 (Synchronization in expectation)**.** *The discrete-time agents* (5.4) *are said to be synchronized in expectation if for all initial states* $\boldsymbol{x}_{0i} \in \mathbb{R}^n$*, the agents satisfy*

$$\lim_{k\to\infty} \|\mathrm{E}(\boldsymbol{x}_i(k) - \boldsymbol{x}_j(k))\| = 0, \quad i, j = 1, 2, \ldots, N. \tag{5.44}$$

5.5.2 Synchronization condition

In this section, a necessary and sufficient synchronization condition for randomly coupled agents is derived. In contrast to the results in Section 5.4, asymptotic synchronization of the expected value of the agent states as defined in (5.44) is considered. The proof of the theorem is given in the appendix.

Theorem 5.3. *The overall system (5.14), which consists of the agents (5.12) and the control algorithm* (5.2) *is synchronized in expectation if and only if*

$$\bar{\lambda}_2\,\rho\left(\boldsymbol{A}^n\right) < 1 \tag{5.45}$$

holds. $\bar{\lambda}_2$ *is the second largest eigenvalue of* $\bar{\boldsymbol{T}}$ *defined in* (5.43).

The result in Theorem 5.3 is similar to the result obtained in Section 5.4.2. Synchronization in expectation can only be achieved if (5.45) holds. Since $\bar{\boldsymbol{T}}$ is a symmetric doubly stochastic matrix, it is easy to verify that the eigenvalues of $\bar{\boldsymbol{T}}$ are less than or equal to one: $1 = \bar{\lambda}_1 \geq \bar{\lambda}_2 \geq \ldots \geq \bar{\lambda}_N$. Furthermore, following from the definition in (5.43), $\bar{\lambda}_2 < 1$ holds if the graph that corresponds to the Laplacian matrix $\boldsymbol{L} = \boldsymbol{L}_1 + \boldsymbol{L}_2 + \ldots + \boldsymbol{L}_R$ is connected. Hence, semi-stable systems $\left(\rho(\boldsymbol{A}^{Pn}) \leq 1\right)$ are synchronized with respect to the requirement (5.44) if and only if the random couplings of the agents ensure that each agent shares information with every other agent directly or indirectly $\left(\bar{\lambda}_2 < 1\right)$.

Synchronization of unstable agents ($\rho(\boldsymbol{A}) > 1$) can only be achieved if $\bar{\lambda}_2 < 1/\rho\left(\boldsymbol{A}^n\right)$. This equation shows that the synchronization condition (5.45) is influenced by the largest magnitude of the agent eigenvalues and the second largest eigenvalue of $\bar{\boldsymbol{T}}$, which is determined by the structure of the communication network. Since the synchronization conditions (5.26) and (5.45) are very similar, the influence of the spectral radius $\rho(\boldsymbol{A}^n)$ on both conditions is similar too (relating properties can be found in Section 5.4.2).

Equations (2.14), (5.42) and the following computation

$$\begin{aligned}\lim_{r\to\infty} \mathrm{E}(\boldsymbol{x}(rn)) &= \left(\lim_{r\to\infty} \bar{\boldsymbol{T}}^r \otimes \lim_{r\to\infty} \boldsymbol{A}^{rn}\right) \boldsymbol{x}_0 \\ &= \left(\frac{1}{N} \mathbb{1}\mathbb{1}^T \otimes \lim_{r\to\infty} \boldsymbol{A}^{rn}\right) \boldsymbol{x}_0 \\ &= \left(\mathbb{1} \otimes \lim_{r\to\infty} \boldsymbol{A}^{rn} \frac{1}{N} \sum_{i=1}^{N} \boldsymbol{x}_{0i}\right)\end{aligned}$$

show that the expected value of the asymptotic agent behavior meets the synchronous trajectory given in eqn. (5.5).

5.5.3 Design of the sampling time to guarantee performance

Section 5.4.3 shows how the sampling time T of deterministically coupled agents can be designed in order to satisfy requirement (5.3). Since the effect of the sampling time on the synchronization of randomly coupled agents is very similar compared to deterministically coupled agents, the same approach is used in this section for the determination of the sampling time T to guarantee additional performance requirements. However, there are differences in the analysis of randomly coupled agents, since the expected behavior of the agents is considered. The sampling time T is designed to satisfy a bound on the expected behavior of the sum of synchronization errors given by

$$\left(\sum_{i=2}^{N} \|\mathrm{E}\left(\boldsymbol{x}_1(t) - \boldsymbol{x}_i(t)\right)\|^2\right)^{\frac{1}{2}} \leq 0.05 \,\|\boldsymbol{e}_0\|, \quad \text{for } t \geq \tilde{T}_{5\%}. \tag{5.46}$$

The time $\tilde{T}_{5\%}$ represents the required synchronization time.

The procedure for the design of the sampling time T is motivated by the results of Section 5.4.3. First, the estimation of the number of time steps $\tilde{k}_{5\%}$ is made, after which

$$\left(\sum_{i=2}^{N} \|\mathrm{E}\left(\boldsymbol{x}_1(k) - \boldsymbol{x}_i(k)\right)\|^2\right)^{\frac{1}{2}} \leq 0.05\,\|\boldsymbol{e}_0\|, \quad \text{for } k \geq \tilde{k}_{5\%} \tag{5.47}$$

is achieved. This result is stated in Lemma 5.2. Secondly, the estimation is used for the proof of Theorem 5.4 in which the sampling time T is determined to satisfy (5.46).

The overall representation of the sum of expected synchronization errors in (5.47) is obtained by

$$\begin{aligned}\left(\sum_{i=2}^{N} \|\underbrace{\mathrm{E}\left(\boldsymbol{x}_1(t) - \boldsymbol{x}_i(t)\right)}_{\mathrm{E}\left(\boldsymbol{e}_{i-1}(k)\right)}\|^2\right)^{\frac{1}{2}} &= \left\| \begin{matrix} \mathrm{E}\left(\boldsymbol{e}_1(k)\right) \\ \vdots \\ \mathrm{E}\left(\boldsymbol{e}_{N-1}(k)\right) \end{matrix} \right\| \\ &= \|\underbrace{\left(\boldsymbol{N} \otimes \boldsymbol{I}\right) \mathrm{E}\left(\boldsymbol{x}(k)\right)}_{\tilde{\boldsymbol{e}}(k)}\|. \end{aligned} \tag{5.48}$$

The corresponding behavior of the expected synchronization error

$$\tilde{\mathbf{e}}(rn) = \left(\boldsymbol{N} \otimes \boldsymbol{I}\right)\left(\bar{\boldsymbol{T}} \otimes \boldsymbol{A}^n\right)^r \boldsymbol{x}_0, \quad r = 0, 1, 2, \ldots \tag{5.49}$$

results from eqn. (5.42). Since $\bar{\boldsymbol{T}}$ is a symmetric doubly stochastic matrix, it is easy to show that the similarity transformation with the involutory matrix $\boldsymbol{V}$ given in (5.29) leads to a decomposition of the matrix $\bar{\boldsymbol{T}}$ into an upper block triangular matrix of the form

$$\boldsymbol{V}^{-1}\bar{\boldsymbol{T}}\boldsymbol{V} = \left(\begin{array}{c|ccc} 1 & * & \cdots & * \\ \hline 0 & & & \\ \vdots & & \tilde{\bar{\boldsymbol{T}}} & \\ 0 & & & \end{array}\right) \tag{5.50}$$

with

$$\rho\left(\tilde{\bar{\boldsymbol{T}}}\right) = \bar{\lambda}_2. \tag{5.51}$$

The application of the transformation (5.50) to (5.49) shows that the behavior of the expected synchronization error $\tilde{e}(k)$ is given by

$$\begin{aligned}
\tilde{e}(rn) &= (\boldsymbol{N} \otimes \boldsymbol{I})\left(\bar{\boldsymbol{T}}^r \otimes \boldsymbol{A}^{rn}\right)\boldsymbol{x}_0 \\
&= (\boldsymbol{N} \otimes \boldsymbol{I})\left(\boldsymbol{V}^2\,\bar{\boldsymbol{T}}^r\,\boldsymbol{V}^2 \otimes \boldsymbol{A}^{rn}\right)\boldsymbol{x}_0 \\
&= (\boldsymbol{N}\boldsymbol{V} \otimes \boldsymbol{I})\left(\boldsymbol{V}\,\bar{\boldsymbol{T}}\,\boldsymbol{V} \otimes \boldsymbol{A}^{n}\right)^r(\boldsymbol{V} \otimes \boldsymbol{I})\,\boldsymbol{x}_0 \\
&= \left(\left(\boldsymbol{0}\,|\,\boldsymbol{I}\,\right)\otimes \boldsymbol{I}\right)\left(\left(\begin{array}{c|ccc} 1 & * & \cdots & * \\ \hline 0 & & & \\ \vdots & & \bar{\tilde{\boldsymbol{T}}}^r & \\ 0 & & & \end{array}\right)\otimes \boldsymbol{A}^{rn}\right)(\boldsymbol{V} \otimes \boldsymbol{I})\,\boldsymbol{x}_0 \\
&= \left(\bar{\tilde{\boldsymbol{T}}}^r \otimes \boldsymbol{A}^{rn}\right)\left(\left(\boldsymbol{0}\,|\,\boldsymbol{I}\,\right)\boldsymbol{V} \otimes \boldsymbol{I}\right)\boldsymbol{x}_0 \\
&= \left(\bar{\tilde{\boldsymbol{T}}}^r \otimes \boldsymbol{A}^{rn}\right)\underbrace{(\boldsymbol{N} \otimes \boldsymbol{I})\,\boldsymbol{x}_0}_{\boldsymbol{e}(0)\,=\,\boldsymbol{e}_0}, \qquad r = 0,\,1,\,2,\,\dots\,.
\end{aligned} \tag{5.52}$$

Note that $\boldsymbol{A}^n = \mathrm{e}^{\boldsymbol{A}_\mathrm{c}Tn} = \tilde{\boldsymbol{A}}^T$ and $\tilde{\boldsymbol{A}} = \mathrm{e}^{\boldsymbol{A}_\mathrm{c}n}$ holds. The estimation of the synchronization time $\tilde{k}_{5\%}$ is obtained by considering (5.46), (5.48) and (5.52). The substitution of (5.48) and (5.52) into (5.46) gives the inequality

$$\begin{aligned}
\|\tilde{e}(rn)\| &= \left\|\left(\bar{\tilde{\boldsymbol{T}}}^r \otimes \boldsymbol{A}^{rn}\right)\boldsymbol{e}_0\right\| \\
&= \left\|\left(\bar{\tilde{\boldsymbol{T}}}^r \otimes \tilde{\boldsymbol{A}}^{rT}\right)\boldsymbol{e}_0\right\| \leq 0.05\|\boldsymbol{e}_0\|,
\end{aligned} \tag{5.53}$$

which is satisfied for $r \geq \tilde{r}_{5\%}$ with respect to the similarity transformations $\tilde{\boldsymbol{V}}_\mathrm{b}^{-1}\,\bar{\tilde{\boldsymbol{T}}}\,\tilde{\boldsymbol{V}}_\mathrm{b} = \mathrm{diag}(\lambda_i(\bar{\tilde{\boldsymbol{T}}}))$, $\boldsymbol{V}_\mathrm{a}^{-1}\,\tilde{\boldsymbol{A}}\,\boldsymbol{V}_\mathrm{a} = \mathrm{diag}(\lambda_i(\tilde{\boldsymbol{A}}))$ and eqn. (5.51) if

$$\begin{aligned}
\left\|\left(\bar{\tilde{\boldsymbol{T}}}^r \otimes \tilde{\boldsymbol{A}}^{rT}\right)\boldsymbol{e}_0\right\| &\leq \left\|\bar{\tilde{\boldsymbol{T}}}^r\right\|\left\|\tilde{\boldsymbol{A}}^{rT}\right\|\|\boldsymbol{e}_0\| \\
&\leq \left\|\tilde{\boldsymbol{V}}_\mathrm{b}\right\|\left\|\tilde{\boldsymbol{V}}_\mathrm{b}^{-1}\right\|\underbrace{\rho\left(\bar{\tilde{\boldsymbol{T}}}\right)^r}_{\bar{\lambda}_2^r}\|\boldsymbol{V}_\mathrm{a}\|\left\|\boldsymbol{V}_\mathrm{a}^{-1}\right\|\rho\left(\tilde{\boldsymbol{A}}\right)^{rT}\|\boldsymbol{e}_0\| && (5.54) \\
&\leq \|\boldsymbol{e}_0\|\,0.05 && (5.55)
\end{aligned}$$

holds. As the system matrix is assumed to be semi-stable, it follows from $\rho\left(\tilde{\boldsymbol{A}}\right)^{rT} = 1$ and eqn. (5.55) that the inequality (5.53) is fulfilled for $r \geq \tilde{r}_{5\%}$ if the following is true

$$\left\|\tilde{\boldsymbol{V}}_\mathrm{b}\right\|\left\|\tilde{\boldsymbol{V}}_\mathrm{b}^{-1}\right\|\|\boldsymbol{V}_\mathrm{a}\|\left\|\boldsymbol{V}_\mathrm{a}^{-1}\right\|\bar{\lambda}_2^{\tilde{r}_{5\%}} \leq 0.05. \tag{5.56}$$

Remark 5.4. *The inequality in* (5.54) *shows that for semi-stable agents the convergence rate of the expected synchronization error* $\tilde{\mathrm{e}}(k)$ *is solely a property of the agent coupling, which is expressed by* $\bar{\lambda}_2$.

The following lemma is a direct consequence of (5.56).

Lemma 5.2. *A network of semi-stable agents that fulfills the synchronization condition* (5.45) *is synchronized with respect to inequality* (5.47) *after* $\tilde{k}_{5\%} = n\,\tilde{r}_{5\%}$ *time steps, where*

$$\tilde{r}_{5\%} = \left\lceil \log^{-1}\left(\bar{\lambda}_2\right) \log\left(\frac{0.05}{\|\boldsymbol{V}_{\mathrm{a}}\| \, \|\boldsymbol{V}_{\mathrm{a}}^{-1}\| \, \left\|\tilde{\boldsymbol{V}}_{\mathrm{b}}\right\| \, \left\|\tilde{\boldsymbol{V}}_{\mathrm{b}}^{-1}\right\|}\right)\right\rceil . \tag{5.57}$$

Proof. Inequality (5.47) is equivalent to (5.53), which is fulfilled if (5.56) holds. The lemma is proved by considering eqn. (5.57), which results from (5.45) and the synchronization condition (5.56). The ceiling function in eqn. (5.57) is used to create an integer-valued upper bound. □

Remark 5.5. *The result of Lemma 5.2 shows that after* $\tilde{k}_{5\%}$ *time steps synchronization with respect to* (5.47) *is achieved. Furthermore, since a dead-beat controller is used and synchronization of coupled agents is achieved in* $k = n$ *time steps, there exists a direct connection between the time for synchronization* $\tilde{k}_{5\%}$ *and the associated number of couplings between the agents. According to eqn.* (5.57)*, the number of couplings for synchronization with respect to inequality* (5.47) *is determined by*

$$\tilde{N}_{5\%} = \tilde{r}_{5\%}\, N_{\mathrm{r}} \tag{5.58}$$

where N_{r} *is the maximal number of couplings in a synchronization step. Equation* (5.58) *gives an answer to the question: How many couplings are sufficient for synchronization in expectation of randomly coupled agents?*

The estimation in eqn. (5.57) is elaborated to be independent of the sampling time T. In particular, this is true for semi-stable agents, which can be seen in (5.54) and (5.56). This property allows the determination of the sampling time T from eqn. (5.57) to satisfy requirement (5.46). Thus, by considering the relation between the sampling time T, the discrete-time steps k and the corresponding continuous-time t it is easy to see that synchronization with respect to (5.3) is achieved if

$$\tilde{k}_{5\%}\, T \leq \tilde{T}_{5\%} \tag{5.59}$$

holds. Since the estimated number of the synchronization steps $\tilde{k}_{5\%}$ is independent of the sampling time T, eqn. (5.59) can directly be used for the determination of the sampling time T to satisfy the requirement (5.46) for any given synchronization time $\tilde{T}_{5\%}$. This result is summarized in the following theorem.

Theorem 5.4. *Semi-stable agents that fulfill the synchronization condition* (5.45) *are synchronized with respect to inequality* (5.46) *for the given synchronization time* $\tilde{T}_{5\%}$ *if*

$$T \leq \frac{\tilde{T}_{5\%}}{\tilde{k}_{5\%}}$$

holds. $\tilde{k}_{5\%}$ *is the estimation given in Lemma 5.2.*

5.6 Example: Synchronization of harmonic oscillators

5.6.1 Model of harmonic oscillators

As an example, the synchronization of $N = 4$ harmonic oscillators is considered. Harmonic oscillators have received considerable attention in literature on synchronization (cf. [131], [44], [132]), since they can be used for the modeling of a large variety of technical systems, such as chemical processes, electrical oscillators or even for the modeling of sensors.

The continuous-time model of the agents is given by the second-order state-space systems

$$\dot{\boldsymbol{x}}_i(t) = \underbrace{\begin{pmatrix} 0 & 1 \\ -\omega^2 & -2\,d\,\omega \end{pmatrix}}_{\boldsymbol{A}_{\mathrm{c}}} \boldsymbol{x}_i(t) + \underbrace{\begin{pmatrix} 0 \\ k_{\mathrm{s}}\,\omega^2 \end{pmatrix}}_{\boldsymbol{b}_{\mathrm{c}}} u_i(t), \quad \boldsymbol{x}_i(0) = \boldsymbol{x}_{0i}, \qquad i = 1, 2, 3, 4 \tag{5.60}$$

$$y_i(t) = \underbrace{\begin{pmatrix} 1 & 0 \end{pmatrix}}_{\boldsymbol{c}_{\mathrm{c}}^T} \boldsymbol{x}_i(t) \tag{5.61}$$

with the natural frequency $\omega = 0.5$, the static gain $k_{\mathrm{s}} = 4$ and the damping d. The control algorithms, designed in this chapter, are applied in the following example for the synchronization of harmonic oscillators ($d = 0$) and for unstable systems with $d = -0.06$. The discrete-time model of harmonic oscillators is given by

$$\boldsymbol{x}_i(k+1) = \underbrace{\begin{pmatrix} \cos(\omega T) & \sin(\omega T) \\ -\sin(\omega T) & \cos(\omega T) \end{pmatrix}}_{\boldsymbol{A}} \boldsymbol{x}_i(k) + \underbrace{\begin{pmatrix} 1-\cos(\omega T) \\ \sin(\omega T) \end{pmatrix}}_{\boldsymbol{b}} k_{\rm s}\, u_i(k), \tag{5.62}$$

$$y_i(t) = \underbrace{\begin{pmatrix} 1 & 0 \end{pmatrix}}_{\boldsymbol{c}^T} \boldsymbol{x}_i(k), \quad \boldsymbol{x}_i(0) = \boldsymbol{x}_{0i}, \tag{5.63}$$

where T is sampling time. In the case where $d = -0.06$ the components of the discrete-time state-space model (5.62)–(5.63) are given by

$$\boldsymbol{A} = \mathrm{e}^{\boldsymbol{A}_{\rm c}T}, \quad \boldsymbol{b} = \boldsymbol{A}_{\rm c}^{-1}(\mathrm{e}^{\boldsymbol{A}_{\rm c}T} - \boldsymbol{I})\boldsymbol{b}_{\rm c} \text{ and } \boldsymbol{c}^T = \boldsymbol{c}_{\rm c}^T.$$

The following simulations show the continuous-time behavior of the agents (5.60) – (5.61), which are controlled by the deterministic gossip algorithm of Section 5.4 and the random gossip algorithm of Section 5.5. The initial values are given by

$$\boldsymbol{x}_{01} = \begin{pmatrix} 3 \\ -1 \end{pmatrix}, \; \boldsymbol{x}_{02} = \begin{pmatrix} 0 \\ 1 \end{pmatrix}, \; \boldsymbol{x}_{03} = \begin{pmatrix} 1 \\ 0.2 \end{pmatrix} \text{ and } \boldsymbol{x}_{04} = \begin{pmatrix} -2 \\ 0 \end{pmatrix}.$$

The control vector $\boldsymbol{k}^T$ in eqn. (5.6) is designed according to eqn. (5.9) for the discrete-time agent model (5.62)–(5.63) and the given sampling time T.

First, the synchronization of harmonic oscillators ($d = 0$) is considered for deterministically and for randomly coupled agents. For both coupling methods the sampling time T is designed to fulfill the requirement on the synchronization time $T_{5\%}$ given in (5.3) and (5.46), respectively. The effectiveness of the corresponding design methods is demonstrated in Fig. 5.7 and Fig. 5.8.

Secondly, the validity of the necessary and sufficient synchronization conditions given in Theorem 5.1 and Theorem 5.3 are verified in simulations by considering the synchronization of unstable agents ($d = -0.06$). Figure 5.9 shows the synchronous trajectory, and Fig. 5.10 the synchronization errors $\boldsymbol{e}(k)$ of the unstable agents.

5.6.2 Synchronization of harmonic oscillators

Deterministic agent coupling

The following investigations are based on the coupling structure presented in Example 5.1. For the verification of the synchronization condition (5.26) the second eigenvalue of the matrix $\boldsymbol{T}$

in (5.25) has first to be determined. The deterministic agent coupling shown in Fig. 5.4 is achieved with the period $P = 3$. Hence, the matrix $\boldsymbol{T}$ is given by the matrix product

$$\boldsymbol{T} = \boldsymbol{T}_3 \boldsymbol{T}_2 \boldsymbol{T}_1 = \begin{pmatrix} 0.5 & 0.5 & 0 & 0 \\ 0.125 & 0.125 & 0.375 & 0.375 \\ 0.25 & 0.25 & 0.25 & 0.25 \\ 0.125 & 0.125 & 0.375 & 0.375 \end{pmatrix},$$

where $\boldsymbol{T}_i = \boldsymbol{I} - \frac{1}{2}\boldsymbol{L}_i$, $(i = 1, 2, 3)$ (Fig. 5.5). The second largest magnitude among the eigenvalues of $\boldsymbol{T}$ lies inside the unit circle of the complex plane:

$$|\lambda_2| = 0.25. \tag{5.64}$$

The spectral radius $\rho(\boldsymbol{A}) = 1$, which is for the case of harmonic oscillators independent of the sampling time T, and the second eigenvalue in eqn. (5.64) show that

$$|\lambda_2|\,\rho(\boldsymbol{A}^6) = 0.25 < 1$$

holds and the synchronization condition (5.26) is fulfilled. Consequently, the overall closed-loop system, consisting of the agent dynamics (5.62) – (5.63) and the control algorithm 5.1, synchronizes for arbitrary sampling times T.

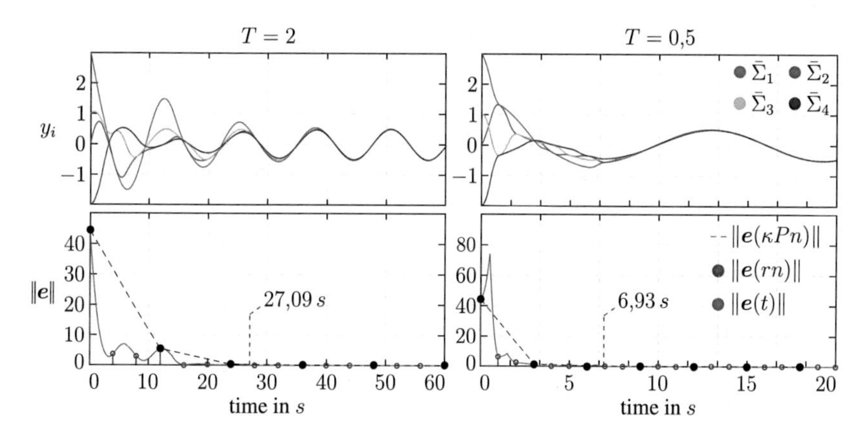

Figure 5.7: Synchronization of deterministically coupled harmonic oscillators

Even if the choice of the sampling time does not affect the synchronization condition (5.26), it has great effects on the convergence rate of the synchronization errors $\boldsymbol{e}_i(t) = \boldsymbol{x}_1(t) - \boldsymbol{x}_{i+1}(t)$, $(i = 1, 2, 3)$ as discussed in Section 5.4.3. Reducing the sampling time T reduces the time for synchronization, as illustrated in Fig. 5.7. This relation is demonstrated by requiring the synchronization error $\boldsymbol{e}(t)$ to be bounded with respect to (5.3) for $T_{5\%} = 60\,s$ and $T_{5\%} = 15\,s$.

According to Lemma 5.1, the number of time steps is estimated that fulfills (5.2):

$$k_{5\%} = 30.$$

The application of Theorem 5.2 shows that synchronization for the required time $T_{5\%} = 60\,s$ and $T_{5\%} = 15\,s$ is achieved for $T \leq 2\,s$ and $T \leq 0.5\,s$, respectively. Figure 5.7 confirms that not only the requirement on asymptotic synchronization of the agents is satisfied but also the requirement on the synchronization time $T_{5\%}$. Furthermore, it is evident from the behavior of the synchronization error in Fig. 5.7 that the harmonic oscillators are already synchronized after $t = 27.09\,s < 60\,s$ and $t = 6.93\,s < 15\,s$. The estimation of the sampling time appears in this example to be two times lower than necessary. In view of Remark 5.3, the number of couplings for synchronization is less than $N_{5\%} = \kappa_{5\%} N_\kappa = 20$, where $N_\kappa = 4$ is the number of agent couplings within one period shown in Fig. 5.4.

Besides the effect of the choice of the sampling time T on the convergence rate of the agents, it is evident from the second column of Fig. 5.7, how the agents behave in relation to the periodically changing interconnection structure described in Example 5.1. In the first step the agent pairs (Σ_1, Σ_2) and (Σ_3, Σ_4) synchronize after $t = nT = 1\,\mathrm{s}$. Then, the agent pairs (Σ_2, Σ_3) and (Σ_2, Σ_4) are synchronized in the second step and the third step. A repetition of the sequence results in the asymptotic synchronization of the agents.

Random agent coupling

The following simulations consider the coupling structure presented in Example 5.2. Figure 5.6 shows the coupling of the agents, where the pairs (Σ_1, Σ_2) and (Σ_3, Σ_4) are coupled simultaneously with probability $p_1 = 14/24$. The coupling of the agent pair (Σ_2, Σ_3) or (Σ_2, Σ_4) is achieved with probability $p_2 = p_3 = 5/24$ (cf. eqn. (5.40) and (5.41)). The synchronization condition in Theorem 5.3 yields

$$\bar{\lambda}_2\, \rho(\boldsymbol{A}^2) = 0.83 < 1, \tag{5.65}$$

where $\bar{\lambda}_2 = 0.83$ is the second largest eigenvalue of the corresponding doubly stochastic matrix

$$\bar{\boldsymbol{T}} = \boldsymbol{I} - \frac{1}{2}\sum_{i=1}^{3} p_i\, \boldsymbol{L}_i = \begin{pmatrix} 0.7 & 0.3 & 0 & 0 \\ 0.3 & 0.5 & 0.1 & 0.1 \\ 0 & 0.1 & 0.6 & 0.3 \\ 0 & 0.1 & 0.3 & 0.6 \end{pmatrix}.$$

Since $\rho(\boldsymbol{A}) = 1$, Algorithm 5.2 ensures synchronization in expectation for any sampling time T.

Figure 5.8 shows the behavior of randomly coupled agents. Comparing the transient behavior of deterministically and randomly coupled agents shows differences in the convergence rate. The coupling sequence of the deterministically coupled agents in Example 5.1 is de-

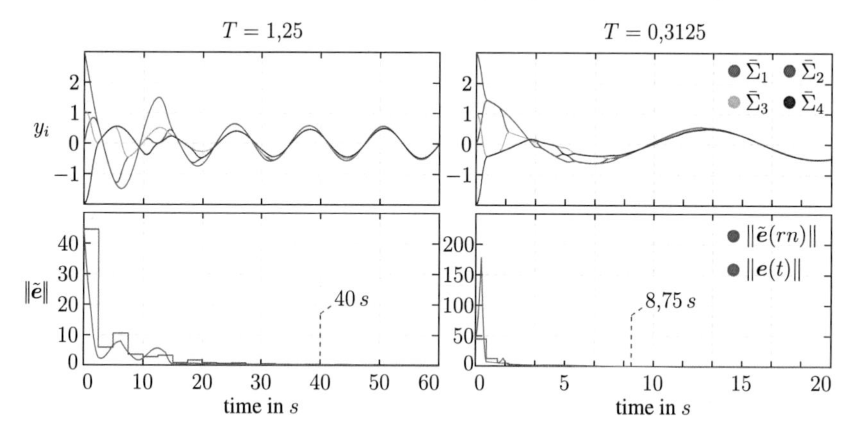

Figure 5.8: Synchronization of randomly coupled harmonic oscillators

signed to establish couplings between agents that are not coupled before. Such a procedure leads intuitively to a faster convergence of the agents towards the synchronous trajectory. In contrast to deterministically coupled agents, a repetitive coupling of the agents appears in the case of random agent couplings. An example for a repetitive coupling can be seen in the right column of Fig. 5.8 for $T = 0.3125\,\mathrm{s}$ and $t \in [0\,\mathrm{s},\, 2T]$. A coupling of the agents (Σ_1, Σ_2) and (Σ_3, Σ_4) is established for $k = 4$ steps, while only $k = n = 2$ steps are necessary for the synchronization in finite time. In the remaining two time steps the synchronization of the overall system stays unaffected. This example shows that for randomly coupled agents there exist situations, in which couplings are introduced that have no effect on the synchronization. Compared to the case of a deterministic agent coupling, repetitive couplings clearly increase the time for synchronization.

The simulations in Fig. 5.8 show the situation where the expected value of the synchronization error $\tilde{e}(t) = \mathrm{E}\left(e(t)\right)$ is required to be bounded with respect to (5.46) for $\tilde{T}_{5\%} = 60\,s$ and $\tilde{T}_{5\%} = 15\,s$. In order to satisfy this requirement, the sampling time T is calculated according to Theorem 5.4. The application of Lemma 5.2 and Theorem 5.4 shows that with

$$\tilde{k}_{5\%} = 48$$

synchronization for the required time $\tilde{T}_{5\%} = 60\,s$ and $\tilde{T}_{5\%} = 15\,s$ is achieved if $T \leq 1.25\,s$ and $T \leq 0.3125\,s$, respectively. Note, that the requirements on the synchronization time are referred to the behavior of the expected synchronization error whereas the simulations only show the agent behavior for one possible coupling sequence.

However, as shown in the second row of Fig. 5.8 the agents are synchronized with respect to the norm of the expected synchronization error after $t = 40\,s < 60\,s$ and $t = 8.75\,s < 15\,s$. The corresponding number of agent couplings that is estimated with eqn. (5.58) is given by $\tilde{N}_{5\%} = \tilde{r}_{5\%}\,N_{\mathrm{r}} = 48$, where $\tilde{r}_{5\%} = \tilde{k}_{5\%}/n = 24$ and $N_{\mathrm{r}} = 2$ holds. Hence, the establishing of $\tilde{N}_{5\%} = 48$ random couplings leads to a synchronization of the agents with respect to (5.47).

5.6.3 Synchronization of unstable oscillators

The following investigation considers the synchronization of the agents (5.60) – (5.61) with $d = -0.06$. The negative damping constant $d = -0.06$ leads to an unstable behavior with the eigenvalues $\lambda_{\mathrm{c}1/2} = 0.03 \pm 0.5\,j$ of the system matrix $\boldsymbol{A}_{\mathrm{c}}$. The corresponding eigenvalues of the discrete-time system are determined by

$$\lambda_i(\boldsymbol{A}) = \mathrm{e}^{\lambda_{\mathrm{c}i}\,T}, \; i = 1,\, 2.$$

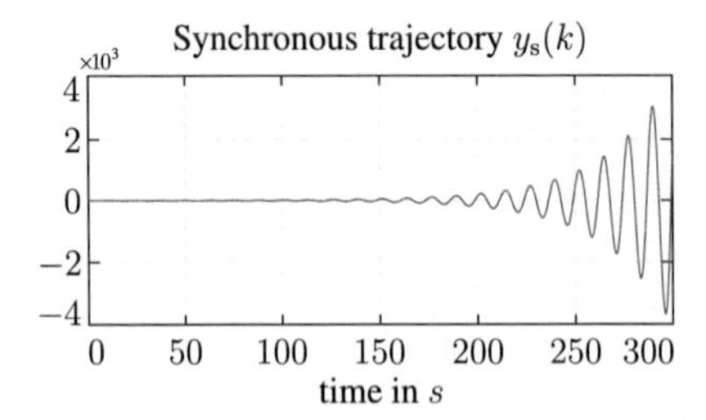

Figure 5.9: Synchronous trajectory of the unstable agents with $d = -0.06$

The communication network used for the simulations is shown in Fig. 5.4 for deterministically and in Fig. 5.6 randomly coupled agents. For a closer examination of the necessary and sufficient synchronization conditions presented in Theorem 5.1 and Theorem 5.3, three simulation scenarios are considered (Fig. 5.10). The first row of Fig. 5.10 shows the simulations for the case where the synchronization conditions are not fulfilled. In the second and the third row of Fig. 5.10 the case of synchronized agents and the limiting case are presented. Figure 5.9 shows the corresponding unstable synchronous trajectory of the agents.

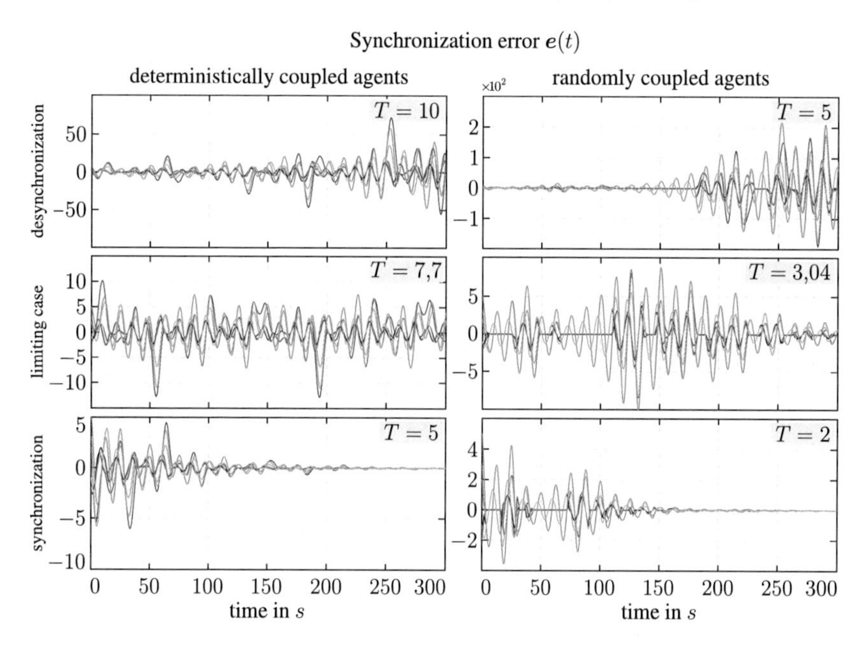

Figure 5.10: Behavior of the synchronization error $\boldsymbol{e}(t)$ for $d = -0.06$ and different sampling times T

Since $\Re(\lambda_{c1/2}) = 0.03$ is greater than zero,

$$\rho(\boldsymbol{A}^{Pn}) = e^{\Re(\lambda_{c1/2})\, P\, n\, T} > 1$$

and

$$\rho(\boldsymbol{A}^{n}) = e^{\Re(\lambda_{c1/2})\, n\, T} > 1$$

holds. Note that for unstable agents $\rho(\boldsymbol{A}^{Pn}) \geq \rho(\boldsymbol{A}^{n})$. This relation shows that the synchronization condition (5.26) for deterministically coupled agents is satisfied if and only if $|\lambda_2| < 1/\rho(\boldsymbol{A}^{Pn}) < 1$ holds. In the case of randomly coupled agents, the synchronization condition (5.45) is satisfied if and only if $\bar{\lambda}_2 < 1/\rho(\boldsymbol{A}^n) < 1$ holds. As already mentioned before, $|\lambda_2|$ and $\bar{\lambda}_2$ are less than one if and only if the union of all agent couplings produces a communication network with a spanning tree. This necessary condition is fulfilled since eqn. (5.64) and (5.65) hold. In order to satisfy the necessary and sufficient synchronization condition (5.26) for deterministically coupled agents, the sampling time T has to be chosen such that

$$|\lambda_2|\rho(\boldsymbol{A}^{Pn}) < 1,$$

which leads to

$$T < \frac{-\ln(|\lambda_2|)}{\Re(\lambda_{c1/2})Pn} = 7.7.$$

In the case of randomly coupled agents the sampling time T has to satisfy

$$T < \frac{-\ln(\bar{\lambda}_2)}{\Re(\lambda_{c1/2})n} = 3.04$$

for ensuring synchronization in expectation.

The synchronization error $e(t)$ in Fig. 5.10 confirms the validity of the synchronization conditions (5.26) and (5.45). Since synchronization in expectation is considered in the synchronization condition (5.45), the simulation belonging to the synchronization of randomly coupled agents shows only a sample of possible agent couplings.

In contrast to the behavior of randomly coupled agents, deterministically coupled agents attain synchronization for a larger sampling time T. This observation confirms the observation of Section 5.6.2, where it is shown in the simulations that couplings appear in the case of randomly coupled agents which have no effect on the synchronization of the overall system. Hence, synchronization of deterministically coupled agents is achieved for a lower number of agent couplings compared to the case of a random agent coupling, which is also shown by $N_{5\%} \leq \tilde{N}_{5\%}$.

Publications

The results concerning the deterministic agent coupling have been published in [6].

6 Conclusions

This thesis has presented three different design approaches for the synchronization of networked multi-agent systems. The main focus was on the performance of the overall system which was described by different types of performance specifications depending on the needs of the specific application. Furthermore, the proposed controller design methods differ with respect to the locally available coupling signals. The coupling signals are specified by the output information of the agents and the characteristics of the communication network, which is assumed to be undirected for the design methods presented in Chapter 3 and 5 or directed as in the extensions section of the design method presented in Chapter 4.

The application of LQR methods for achieving optimal synchronization in the sense of minimizing an LQ-like performance index (Chapter 3) shows that a direct solution of the corresponding optimization problem for synchronization is computationally expensive to obtain. For this reason a gradient descent algorithm is proposed that ensures optimal synchronization for any given static networked controller satisfying the requirement on asymptotic synchronization (Algorithm 3.1). A computationally less expensive design procedure was derived in Section 3.7. In this case, the optimal solution was obtained by a decomposition of the overall problem into subproblems of the order of a single agent and it has been shown that an all-to-all coupling between the agents is required for optimal synchronization. For communication networks which do not couple all agents with all others an approximate networked controller has been derived which synchronizes the agents according to the topology of the given communication network (Section 3.8). For the special case of a complete communication network it is shown that the approximate networked controller is also optimal.

The design of dynamic networked controllers, which solve the output synchronization problem of single-input/single-output agents that are affected by disturbances has been addressed in Chapter 4. The settling time of some overall synchronization error has been indicated to serve as a reasonable quantitative performance specification. Additionally, it has been shown that the damping can also be used as a tuning factor to shape the quality of the transient behavior of the synchronization error. A root locus-based method has been derived, which allows for guaranteeing a specified settling time as well as a specified minimum damping. The approach offers formulas to directly calculate the controller parameters for the given performance requirements. Moreover, it provides a depth graphical insight into the impact of both, the local dynamics of the controlled agents and the network, on the transient behavior of the multi-agent system via the root locus plot. Furthermore, the design approach was extended to cover the synchronization problem of non-identical dynamics and directed communication networks.

For communication networks with the assumption that each agent can only communicate with one other agent at the same time a control algorithm was presented which consists of two decentralized components, the communication unit and the synchronizing controller. The communication unit is used to establish couplings to neighboring agents. For a pair of coupled agents a dead-beat controller ensures synchronization in finite time. After the synchronization of a coupled pair of agents is achieved, the communication unit is allowed to establish a coupling to exactly one other neighbor, which is afterwards used for synchronization. Two types of agent couplings have been considered, a deterministic and a random coupling. The deterministic coupling of the agents is achieved with a periodic interconnection sequence. The behavior of such networked agents has been analyzed and equations have been derived that are suitable to investigate conditions for asymptotic synchronization. The main contributions belonging to this part of the thesis are the necessary and sufficient synchronization conditions, which show that asymptotic synchronization of unstable agents can only be achieved if the synchronization of coupled pairs of agents is reached fast enough.

The design methods have been compared in experiments in the formation control of mobile robots and in simulations, where the synchronization of harmonic oscillators was considered.

Bibliography

Contributions of the author

[1] A. Mosebach and J. Lunze. "Optimal synchronization of circulant networked multi-agent systems". In: *Proc. of the European Control Conference*. 2013, pp. 3815–3820.

[2] A. Mosebach and J. Lunze. "Synchronization of multi-agent systems with similar dynamics". In: *Proc. of the Workshop on Distributed Estimation and Control in Networked Systems*. Vol. 46. 27. 2013, pp. 102–109.

[3] A. Mosebach and J. Lunze. "Synchronization of autonomous agents by an optimal networked controller". In: *Proc. of the European Control Conference*. 2014, pp. 208–213.

[4] A. Mosebach and J. Lunze. "LQR design of synchronizing controllers for networked multi-agent systems". In: *Automatisierungstechnik* 63.6 (2015), pp. 403–412.

[5] A. Mosebach, J. Lunze, and C. Kampmeyer. "Transient behavior of synchronized agents: A design method for dynamic networked controllers". In: *Proc. of the European Control Conference*. 2015.

[6] A. Mosebach and J. Lunze. "A deterministic gossiping algorithm for the synchronization of multi-agent systems". In: *Proc. of the Workshop on Distributed Estimation and Control in Networked Systems*. Vol. 48. 22. 2015, pp. 1–7.

[7] A. Mosebach, S. Röchner, and J. Lunze. "Koordination vernetzter Fahrzeuge an einer Fahrbahnzusammenführung". In: *Automatisierungstechnik* 64.1 (2016), pp. 51–60.

[8] A. Mosebach, S. Röchner, and J. Lunze. "Merging control of cooperative vehicles". In: *Proc. of the Symposium on Advances in Automotive Control*. Vol. 49. 11. 2016, pp. 168–174.

Supervised theses

[9] B. Kilic. *Modellbildung eines mobilen Roboters und Erstellung eines 3D-Modells mit der Simulink 3D-Animation-Toolbox*. Bachelor's thesis, Ruhr-University Bochum, Institute of automation and computer control, 2012.

[10] T. Köllner. *Simulative Untersuchung des Synchronisationverhaltens autonomer Agenten durch vernetzte Regelung*. Bachelor's thesis, Ruhr-University Bochum, Institute of automation and computer control, 2012.

[11] P. Geisler. *Optimale Topologie für die Synchronisation von Multiagentensystemen*. Bachelor's thesis, Ruhr-University Bochum, Institute of automation and computer control, 2013.

[12] E. Polyakov. *Flachheitsbasierte Trajektorienfolgeregelung für einen mobilen Roboter*. Master's thesis, Ruhr-University Bochum, Institute of automation and computer control, 2013.

[13] Y. Wang. *Entwurf synchronisierender Regelungen mit zyklischer Kommunikationstopologie*. Master's thesis, Ruhr-University Bochum, Institute of automation and computer control, 2013.

[14] G. Goldbach. *Entwurf eines Beobachters für die Synchronisation ähnlicher Teilsysteme*. Master's thesis, Ruhr-University Bochum, Institute of automation and computer control, 2014.

[15] F. Just. *String Stability of Digitally Networked Platooning Systems*. Master's thesis, Ruhr-University Bochum, Institute of automation and computer control, 2014.

[16] C. Kampmeyer. *Analysis and Design of Synchronizing Controllers Using Frequency Domain Methods*. Master's thesis, Ruhr-University Bochum, Institute of automation and computer control, 2014.

[17] D. Michaelis. *Synchronisation vernetzter Agenten nach Ausfall von Informationskopplungen*. Master's thesis, Ruhr-University Bochum, Institute of automation and computer control, 2014.

[18] S. Mozian. *Entwurf und Implementierung einer Regelung zur Formation mobiler Roboter*. Bachelor's thesis, Ruhr-University Bochum, Institute of automation and computer control, 2014.

[19] S. Röchner. *Koordination vernetzter Fahrzeuge an einem Einfädelpunkt*. Master's thesis, Ruhr-University Bochum, Institute of automation and computer control, 2014.

[20] Y. Erkul. *Erprobung von Verfahren zur optimalen Synchronisation von Multiagentensystemen*. Master's thesis, Ruhr-University Bochum, Institute of automation and computer control, 2015.

[21] M. Quast. *Synchronisation von Multiagentensystemen mit reduziertem Kommunikationsaufwand*. Master's thesis, Ruhr-University Bochum, Institute of automation and computer control, 2015.

[22] A. Sonders. *Synchronisation von Multiagentystemen in Kommunikationsnetzen mit Punkt-zu-Punkt-Verbindungenens*. Master's thesis, Ruhr-University Bochum, Institute of automation and computer control, 2015.

[23] S. Thaleiser. *Flachheitsbasierte Regelung: Ein Ansatz zur kooperativen Trajektorienplanung mobiler Roboter*. Bachelor's thesis, Ruhr-University Bochum, Institute of automation and computer control, 2015.

Further literature

[24] Y. Cao, W. Yu, W. Ren, and G. Chen. "An overview of recent progress in the study of distributed multi-agent coordination". In: *IEEE Transactions on Industrial Informatics* 9.1 (2013), pp. 427–438.

[25] W. Ren, R.W. Beard, and E.M. Atkins. "A survey of consensus problems in multi-agent coordination". In: *Proc. of the American Control Conference*. 2005, pp. 1859 –1864.

[26] R.O. Saber, J.A. Fax, and R.M. Murray. "Consensus and cooperation in networked multi-agent systems". In: *Proceedings of the IEEE* 95.1 (2007), pp. 215–233.

[27] S. Khodaverdian and J. Adamy. "Distributed dynamic decoupling-based output synchronization for networks of linear heterogeneous MIMO agents". In: *Proc. of the Conference on Decision and Control*. 2015, pp. 6202–6208.

[28] J. Strubel, G.L. Stein, and U. Konigorski. "Synchronization of heterogeneous agents using min-max optimization". In: *Proc. of the American Control Conference*. 2015, pp. 50–55.

[29] G. Seyboth, W. Ren, and F. Allgöwer. "Cooperative control of linear multi-agent systems via distributed output regulation and transient synchronization". In: *Automatica* 68 (2016), pp. 132 –139.

[30] L.M. Pecora and T.L. Carroll. "Synchronization in chaotic systems". In: *Physical Review Letters* 64.8 (1990), pp. 821–824.

[31] J.K. Hale. "Diffusive coupling, dissipation, and synchronization". In: *Journal of Dynamics and Differential Equations* 9.1 (1997), pp. 1–52.

[32] L.M. Pecora and T.L. Carroll. "Master stability functions for synchronized coupled systems". In: *Physical Review Letters* 80.10 (1998), pp. 2109–2112.

[33] M. Bennett, M.F. Schatz, H. Rockwood, and K. Wiesenfeld. "Huygens's clocks". In: *Proceedings of the Royal Society of London A: Mathematical, Physical and Engineering Sciences* 458.2019 (2002), pp. 563–579.

[34] S.H. Strogatz and I. Stewart. "Coupled oscillators and biological synchronization". In: *Scientific American Magazine* 269 (1993), pp. 68–68.

[35] M. Rosenblum and A. Pikovsky. "Synchronization: From pendulum clocks to chaotic lasers and chemical oscillators". In: *Contemporary Physics* 44.5 (2003), pp. 401–416.

[36] G.V. Osipov, J. Kurths, and C. Zhou. *Synchronization in Oscillatory Networks*. Springer Verlag, 2007.

[37] R.O. Saber and R.M. Murray. "Consensus protocols for networks of dynamic agents". In: *Proc. of the American Control Conference*. Vol. 2. 2003, pp. 951–956.

[38] J.A. Fax and R.M. Murray. "Information flow and cooperative control of vehicle formations". In: *IEEE Transactions on Automatic Control* 49.9 (2004), pp. 1465–1476.

[39] L. Moreau. "Stability of continuous-time distributed consensus algorithms". In: *Proc. of the Conference on Decision and Control*. Vol. 4. 2004, pp. 3998–4003.

[40] W. Ren and R.W. Beard. "Consensus seeking in multiagent systems under dynamically changing interaction topologies". In: *IEEE Transactions on Automatic Control* 50.5 (2005), pp. 655–661.

[41] R.O. Saber and R.M. Murray. "Consensus problems in networks of agents with switching topology and time-delays". In: *IEEE Transactions on Automatic Control* 49.9 (2004), pp. 1520–1533.

[42] P. Lin, Y. Jia, J. Du, and S. Yuan. "Distributed consensus control for networks of second-order agents with switching topology and time-delay". In: *Proc. of the European Control Conference*. 2007, pp. 2841–2846.

[43] J. Qin, C. Yu, and H. Gao. "Coordination for linear multiagent systems with dynamic interaction topology in the leader-following framework". In: *IEEE Transactions on Industrial Electronics* 61.5 (2014), pp. 2412–2422.

[44] L. Scardovi and R. Sepulchre. "Synchronization in networks of identical linear systems". In: *Automatica* 45.11 (2009), pp. 2557–2562.

[45] S.E. Tuna. "Synchronizing linear systems via partial-state coupling". In: *Automatica* 44.8 (2008), pp. 2179–2184.

[46] Z. Li, Z. Duan, G. Chen, and L. Huang. "Consensus of multiagent systems and synchronization of complex networks: A unified viewpoint". In: *IEEE Transactions on Circuits and Systems* 57.1 (2010), pp. 213–224.

[47] H. Zhang, F.L. Lewis, and A. Das. "Optimal design for synchronization of cooperative systems: state feedback, observer and output feedback". In: *IEEE Transactions on Automatic Control* 56.8 (2011), pp. 1948–1952.

[48] K.D. Listmann, J. Adamy, and L. Scardovi. "Synchronisierung identischer linearer Systeme – ein Zugang über LMIs". In: *Automatisierungstechnik* 59.9 (2011), pp. 563–573.

[49] K. Hengster-Movric and F.L. Lewis. "Cooperative optimal control for multi-agent systems on directed graph topologies". In: *IEEE Transactions on Automatic Control* 59.3 (2014), pp. 769–774.

[50] Z. Li, Z. Duan, and L. Huang. "Leader-follower consensus of multi-agent systems". In: *American Control Conference*. 2009, pp. 3256–3261.

[51] W. Ni and D. Cheng. "Leader-following consensus of multi-agent systems under fixed and switching topologies". In: *Systems & Control Letters* 59.3 (2010), pp. 209–217.

[52] Z. Li, G. Wen, Z. Duan, and W. Ren. "Designing fully distributed consensus protocols for linear multi-agent systems with directed graphs". In: *IEEE Transactions on Automatic Control* 60.4 (2015), pp. 1152–1157.

[53] P. Wieland, R. Sepulchre, and F. Allgöwer. "An internal model principle is necessary and sufficient for linear output synchronization". In: *Automatica* 47.5 (2011), pp. 1068–1074.

[54] J. Lunze. "Synchronization of heterogeneous agents". In: *IEEE Transactions on Automatic Control* 57.11 (2012), pp. 2885–2890.

[55] J. Lunze. "Synchronisation of systems with individual dynamics by static networked controllers". In: *Asian Journal of Control* 16.2 (2014), pp. 358–369.

[56] P. Wieland and F. Allgöwer. "An internal model principle for consensus in heterogeneous linear multi-agent systems". In: *Proc. of the Workshop on Estimation and Control of Networked Systems* 42.20 (2009), pp. 7–12.

[57] J. Lunze. "Synchronisierbarkeit autonomer Agenten durch vernetzte Regelungen". In: *Automatisierungstechnik* 59.9 (2011), pp. 531–539.

[58] J. Lunze. "An internal-model principle for the synchronisation of autonomous agents with individual dynamics". In: *Proc. of the Conference on Decision and Control*. 2011, pp. 2106–2111.

[59] K.D. Listmann, A. Wahrburg, J. Strubel, J. Adamy, and U. Konigorski. "Partial-state synchronization of linear heterogeneous multi-agent systems". In: *Proc. of the Conference on Decision and Control*. 2011, pp. 3440–3445.

[60] H. Kim, H. Shim, and J.H. Seo. "Output consensus of heterogeneous uncertain linear multi-agent systems". In: *IEEE Transactions on Automatic Control* 56.1 (2011), pp. 200–206.

[61] U. Münz, A. Papachristodoulou, and F. Allgöwer. "Robust consensus controller design for nonlinear relative degree two multi-agent systems with communication constraints". In: *IEEE Transactions on Automatic Control* 56.1 (2011), pp. 145–151.

[62] H.F. Grip, T. Yang, A. Saberi, and A.A. Stoorvogel. "Output synchronization for heterogeneous networks of non-introspective agents". In: *Automatica* 48.10 (2012), pp. 2444 –2453.

[63] H.L. Trentelman, K. Takaba, and N. Monshizadeh. "Robust synchronization of uncertain linear multi-agent systems". In: *IEEE Transactions on Automatic Control* 58.6 (2013), pp. 1511–1523.

[64] A. Isidori, L. Marconi, and G. Casadei. "Robust output synchronization of a network of heterogeneous nonlinear agents via nonlinear regulation theory". In: *IEEE Transactions on Automatic Control* 59.10 (2014), pp. 2680–2691.

[65] Y.G. Sun, L. Wang, and G. Xie. "Average consensus in networks of dynamic agents with switching topologies and multiple time-varying delays". In: *Systems & Control Letters* 57.2 (2008), pp. 175–183.

[66] T. Liu, D.J. Hill, and J. Zhao. "Synchronization of dynamical networks by network control". In: *IEEE Transactions on Automatic Control* 57.6 (2012), pp. 1574–1580.

[67] O. Demir and J. Lunze. "Event-based synchronisation of multi-agent systems". In: *Proc. of the 4th Conference on Analysis and Design of Hybrid Systems* 45.9 (2012), pp. 1–6.

[68] O. Demir and J. Lunze. "Synchronization of multi-agent systems with event-based communication". In: *Automatisierungstechnik* 62.8 (2014), pp. 535–546.

[69] W. Zhu, Z.P. Jiang, and G. Feng. "Event-based consensus of multi-agent systems with general linear models". In: *Automatica* 50.2 (2014), pp. 552–558.

[70] E. Garcia, Y. Cao, and D.W. Casbeer. "Decentralized event-triggered consensus with general linear dynamics". In: *Automatica* 50.10 (2014), pp. 2633–2640.

[71] H. Li, X. Liao, G. Chen, D.J. Hill, Z. Dong, and T. Huang. "Event-triggered asynchronous intermittent communication strategy for synchronization in complex dynamical networks". In: *Neural Networks* 66 (2015), pp. 1–10.

[72] M.Z.Q. Chen, L. Zhang, H. Su, and C. Li. "Event-based synchronisation of linear discrete-time dynamical networks". In: *IET Control Theory Applications* 9.5 (2015), pp. 755–765.

[73] E. Semsar-Kazerooni and K. Khorasani. "Optimal consensus algorithms for cooperative team of agents subject to partial information". In: *Automatica* 44.11 (2008), pp. 2766–2777.

[74] Y. Cao and W. Ren. "LQR-based optimal linear consensus algorithms". In: *Proc. of the American Control Conference*. 2009, pp. 5204–5209.

[75] X. Wu and M.R. Jovanovic. "Sparsity-promoting optimal control of consensus and synchronization networks". In: *Proc. of the American Control Conference*. 2014, pp. 2936–2941.

[76] C. Liu, Z. Duan, G. Chen, and L. Huang. "L2 norm performance index of synchronization and LQR control synthesis of complex networks". In: *Automatica* 45.8 (2009), pp. 1879–1885.

[77] C. Langbort and V. Gupta. "Minimal interconnection topology in distributed control design". In: *SIAM Journal on Control and Optimization* 48.1 (2009), p. 397.

[78] V. Gupta, B. Hassibi, and R.M. Murray. "A sub-optimal algorithm to synthesize control laws for a network of dynamic agents". In: *International Journal of Control* 78.16 (2005), pp. 1302–1313.

[79] J.C. Delvenne, R. Carli, and S. Zampieri. "Optimal strategies in the average consensus problem". In: *Proc. of the Conference on Decision and Control*. 2007, pp. 2498–2503.

[80] R. Carli, A. Chiuso, L. Schenato, and S. Zampieri. "Optimal synchronization for networks of noisy double integrators". In: *IEEE Transactions on Automatic Control* 56.5 (2011), pp. 1146–1152.

[81] Q. Hui. "Optimal distributed linear averaging". In: *Automatica* 47.12 (2011), pp. 2713–2719.

[82] J.A. Rogge, J.A.K. Suykens, and D. Aeyels. "Consensus over ring networks as a quadratic optimal control problem". In: *Proc. of the Symposium on System Structure and Control*. Vol. 43. 21. 2010, pp. 317–323.

[83] D. Zelazo, S. Schuler, and F. Allgöwer. "Cycles and sparse design of consensus networks". In: *Proc. of the Conference on Decision and Control*. 2012, pp. 3808–3813.

[84] S.E. Tuna. *LQR-based coupling gain for synchronization of linear systems*. Tech. rep. arXiv:0801.3390. 2008.

[85] T. Feng, H. Zhang, Y. Luo, and Y. Wang. "Distributed LQR design for multi-agent systems on directed graph topologies". In: *Proc. of the International Joint Conference on Neural Networks*. 2014, pp. 2732–2737.

[86] D.H. Nguyen. "A sub-optimal consensus design for multi-agent systems based on hierarchical {LQR}". In: *Automatica* 55 (2015), pp. 88–94.

[87] H. Zhang, T. Feng, G.H. Yang, and H. Liang. "Distributed cooperative optimal control for multiagent systems on directed graphs: an inverse optimal approach". In: *IEEE Transactions on Cybernetics* 45.7 (2015), pp. 1315–1326.

[88] T. Yang, A.A Stoorvogel, and A. Saberi. "Consensus for multi-agent systems – synchronization and regulation for complex networks". In: *Proc. of the American Control Conference*. 2011, pp. 5312–5317.

[89] E. Peymani, H.F. Grip, A. Saberi, X. Wang, and T.I. Fossen. "H_∞ almost output synchronization for heterogeneous networks of introspective agents under external disturbances". In: *Automatica* 50.4 (2014), pp. 1026–1036.

[90] S. Khodaverdian and J. Adamy. "Synchronizing linear heterogeneous networks by output homogenization". In: 47.3 (2014), pp. 4687–4692.

[91] Y. Hong, X. Wang, and Z.P. Jiang. "Multi-agent coordination with general linear models: A distributed output regulation approach". In: *Proc. of the International Conference on Control and Automation*. 2010, pp. 137–142.

[92] S. Khodaverdian and J. Adamy. "Root locus design for the synchronization of multi-agent systems in general directed networks". In: *Proc. of the Workshop on Distributed Estimation and Control in Networked Systems*. Vol. 48. 22. 2015, pp. 150–155.

[93] A. Fradkov and I. Junussov. "Synchronization of linear object networks by output feedback". In: (2011), pp. 8188–8192.

[94] P.P. Menon and C. Edwards. "Decentralised static output feedback stabilisation and synchronisation of networks". In: *Automatica* 45.12 (2009), pp. 2910–2916.

[95] K. Hengster-Movric, F.L. Lewis, and M. Sebek. "Distributed static output-feedback control for state synchronization in networks of identical LTI systems". In: *Automatica* 53 (2015), pp. 282–290.

[96] N. Chopra. "Output synchronization on strongly connected graphs". In: *IEEE Transactions on Automatic Control* 57.11 (2012), pp. 2896–2901.

[97] P. Massioni and M. Verhaegen. "Distributed control for identical dynamically coupled systems: a decomposition approach". In: *IEEE Transactions on Automatic Control* 54.1 (2009), pp. 124–135.

[98] Y. Liu and J. Lunze. "Synchronization of heterogeneous leader-follower multi-agent system with external disturbances". In: 4.1 (2013), pp. 397–403.

[99] H. Kim, H. Shim, J. Back, and J.H. Seo. "Consensus of output-coupled linear multi-agent systems under fast switching network: Averaging approach". In: *Automatica* 49.1 (2013), pp. 267 –272.

[100] J. Qin and C. Yu. "Exponential consensus of general linear multi-agent systems under directed dynamic topology". In: *Automatica* 50.9 (2014), pp. 2327–2333.

[101] W. Yu, G. DeLellis P.and Chen, M. Bernardo, and J. Kurths. "Distributed adaptive control of synchronization in complex networks". In: *IEEE Transactions on Automatic Control* 57.8 (2012), pp. 2153–2158.

[102] S. Boyd, A. Ghosh, B. Prabhakar, and D. Shah. "Randomized gossip algorithms". In: *IEEE Transactions on Information Theory* 52.6 (2006), pp. 2508–2530.

[103] J. Liu, S. Mou, S. Morse, B. Anderson, and C. Yu. "Deterministic gossiping". In: *Proceedings of the IEEE* 99.9 (2011), pp. 1505–1524.

[104] G. Shi, M. Johansson, and K.H. Johansson. "Randomized gossiping with unreliable communication: Dependent or independent node updates". In: *51st Conf. on Dec. and Contr.* 2012, pp. 4846–4851.

[105] F. He, S. Mou, J. Liu, and A.S. Morse. "Convergence rate on periodic gossiping". In: *Information Sciences* 364 (2016), pp. 111–125.

[106] S. Mou, C. Yu, B.D.O. Anderson, and A.S. Morse. "Deterministic gossiping with a periodic protocol". In: *Proc. of the Conference on Decision and Control.* 2010, pp. 5787–5791.

[107] C. De Persis, P. Frasca, and J.M. Hendrickx. "Self-triggered rendezvous of gossiping second-order agents". In: *52nd Conf. on Dec. and Contr.* 2013, pp. 7403–7408.

[108] F. Benezit, A.G. Dimakis, P. Thiran, and M. Vetterli. “Order-optimal consensus through randomized path averaging”. In: *IEEE Trans. on Inf. Theor.* 56.10 (2010), pp. 5150–5167.

[109] A.G. Dimakis, S. Kar, J.M.F. Moura, M.G. Rabbat, and A. Scaglione. “Gossip algorithms for distributed signal processing”. In: *Proc. of the IEEE* 98.11 (2010), pp. 1847–1864.

[110] J. Liu, B.D.O. Anderson, M. Cao, and S. Morse. “Analysis of accelerated gossip algorithms”. In: *Automatica* 49.4 (2013), pp. 873–883.

[111] O. Mangoubi, S. Mou, J. Liu, and S. Morse. “Towards optimal convex combination rules for gossiping”. In: *Amer. Contr. Conf.* 2013, pp. 1261–1265.

[112] M. Fiedler. “Algebraic connectivity of graphs”. In: *Czechoslovak mathematical journal* 23.2 (1973), pp. 298–305.

[113] J. Brewer. “Kronecker products and matrix calculus in system theory”. In: *IEEE Transactions on Circuits and Systems* 25.9 (1978), pp. 772–781.

[114] A.J. Laub. *Matrix Analysis for Scientists and Engineers*. Society for Industrial and Applied Mathematics, 2004.

[115] W.H. Steeb and T.K. Shi. *Matrix Calculus and the Kronecker Product with Applications and C++ Programs*. World Scientific, 1997.

[116] R. Merris. “Laplacian matrices of graphs: a survey”. In: *Linear Algebra and its Applications* 197 (1994), pp. 143–176.

[117] B. Mohar, Y. Alavi, G. Chartrand, O.R. Oellermann, and Schwenk A.J. “The Laplacian spectrum of graphs”. In: *Graph Theory, Combinatorics, and Applications* 12.2 (1991), pp. 871–898.

[118] R. Merris. “Doubly stochastic graph matrices, II”. In: *Linear and Multilinear Algebra* 45.2 (1998), pp. 275–285.

[119] K. Zhou, J.C. Doyle, and K. Glover. *Robust and Optimal Control*. Prentice-Hall, 1996.

[120] J. Lunze. *Regelungstechnik 2: Mehrgrößensysteme, Digitale Regelung*. Vol. 8. Springer, 2014.

[121] J. Lunze. *Regelungstechnik 1: Systemtheoretische Grundlagen, Analyse und Entwurf einschleifiger Regelungen*. Vol. 10. Springer, 2014.

[122] K. Ogata. *Modern Control Engineering*. Vol. 5. Prentice Hall, 2009.

[123] A.D.G. Dimakis, A.D. Sarwate, and M.J. Wainwright. "Geographic gossip: efficient averaging for sensor networks". In: *IEEE Transactions on Signal Processing* 56.3 (2008), pp. 1205–1216.

[124] P. Wieland, J.S. Kim, and F. Allgöwer. "On topology and dynamics of consensus among linear high-order agents". In: *International Journal of Systems Science* 42.10 (2011), pp. 1831–1842.

[125] I. Markovsky and S. Van Huffel. "Overview of total least-squares methods". In: *Signal processing* 87.10 (2007), pp. 2283–2302.

[126] W. Levine and M. Athans. "On the determination of the optimal constant output feedback gains for linear multivariable systems". In: *IEEE Transactions on Automatic Control* 15.1 (1970), pp. 44–48.

[127] S. Boyd, L. El-Ghaoui, E. Feron, V. Balakrishnan, and E. Yaz. "Linear matrix inequalities in system and control theory". In: *Proceedings of the IEEE* 85.4 (1997), pp. 698–699.

[128] J. Lunze. "Transient behaviour of synchronised agents". In: *Automatisierungstechnik* 60.7 (2012), pp. 398–404.

[129] J. Lunze. "Finite-time synchronisation of completely coupled agents". In: *Proc. of the Workshop on Distributed Estimation and Control in Networked Systems*. 2013, pp. 316–321.

[130] E.D. Sontag. *Mathematical Control Theory: Deterministic Finite Dimensional Systems*. Vol. 6. Springer Science & Business Media, 2013.

[131] W. Ren. "Synchronization of coupled harmonic oscillators with local interaction". In: *Automatica* 44.12 (2008), pp. 3195–3200.

[132] J. Lunze. "Complete synchronisation of linear oscillator networks". In: *International Journal of Control* 85.7 (2012), pp. 800–814.

Appendices

A Proofs of Chapter 3

A.1 Proof of Theorem 3.1

Proof. Using the state transformation

$$\tilde{\boldsymbol{x}}(t) = (\boldsymbol{T} \otimes \boldsymbol{I})^{-1} \boldsymbol{x}(t) \tag{A.1}$$

with $\boldsymbol{T}^{-1} \boldsymbol{L} \boldsymbol{T} = \text{diag}(\lambda_i)$ the solution (3.10) of the overall closed-loop system (3.9) gets the following form:

$$\begin{aligned}
\tilde{\boldsymbol{x}}(t) &= (\boldsymbol{T} \otimes \boldsymbol{I})^{-1} \mathrm{e}^{(\boldsymbol{I} \otimes \boldsymbol{A} - \boldsymbol{L} \otimes \boldsymbol{B}\boldsymbol{K})\, t}\, (\boldsymbol{T} \otimes \boldsymbol{I})\, (\boldsymbol{T} \otimes \boldsymbol{I})^{-1} \boldsymbol{x}_0 & \text{(A.2)}\\
&= \mathrm{e}^{(\boldsymbol{T} \otimes \boldsymbol{I})^{-1} (\boldsymbol{I} \otimes \boldsymbol{A} - \boldsymbol{L} \otimes \boldsymbol{B}\boldsymbol{K})\, (\boldsymbol{T} \otimes \boldsymbol{I})\, t}\, (\boldsymbol{T} \otimes \boldsymbol{I})^{-1} \boldsymbol{x}_0 \\
&= \mathrm{e}^{\left(\boldsymbol{I} \otimes \boldsymbol{A} - \boldsymbol{T}^{-1} \boldsymbol{L} \boldsymbol{T} \otimes \boldsymbol{B}\boldsymbol{K}\right) t}\, \tilde{\boldsymbol{x}}(0) \\
&= \mathrm{e}^{(\boldsymbol{I} \otimes \boldsymbol{A} - \text{diag}(\lambda_i) \otimes \boldsymbol{B}\boldsymbol{K})\, t}\, \tilde{\boldsymbol{x}}(0) \\
&= \text{diag}\left(\mathrm{e}^{(\boldsymbol{A} - \lambda_i \boldsymbol{B}\boldsymbol{K})\, t}\right) \tilde{\boldsymbol{x}}(0). & \text{(A.3)}
\end{aligned}$$

According to requirement (3.1), the agents are synchronized if and only if every entry of the error vector

$$\boldsymbol{e}(t) = \left(\underbrace{\begin{pmatrix} 1 & -1 & 0 & \cdots & 0 \\ 1 & 0 & -1 & \ddots & \vdots \\ \vdots & \vdots & \ddots & \ddots & 0 \\ 1 & 0 & \cdots & 0 & -1 \end{pmatrix}}_{\boldsymbol{N}} \otimes \boldsymbol{I} \right) \boldsymbol{x}(t) \tag{A.4}$$

or any invertible linear transformation of this vector converges asymptotically to zero. $\boldsymbol{N}$ is a $(N-1 \times N)$-matrix with $\text{rank}(\boldsymbol{N}) = N - 1$ which by construction has a row sum equal to zero. The first eigenvector of the Laplacian matrix $\boldsymbol{L}$ shows that the relation

$$\boldsymbol{N}\boldsymbol{T} = \boldsymbol{N}\left(\frac{1}{\sqrt{N}}\mathbb{1}\ \boldsymbol{t}_2\ \cdots\ \boldsymbol{t}_N\right) = \left(\ 0 \mid \hat{\boldsymbol{N}}\ \right) \tag{A.5}$$

holds. Since $\operatorname{rank}(\boldsymbol{T}) = N$, it holds that $\operatorname{rank}(\boldsymbol{N}\boldsymbol{T}) = \operatorname{rank}(\boldsymbol{N}) = N-1$. Hence, $\hat{\boldsymbol{N}}$ is an invertible $(N-1 \times N-1)$-matrix. The application of the transformation (A.1) and eqns. (A.3) and (A.5) to the error vector (A.4) yields

$$\begin{aligned}
\boldsymbol{e}(t) &= (\boldsymbol{N}\otimes\boldsymbol{I})\ (\boldsymbol{T}\otimes\boldsymbol{I})\ (\boldsymbol{T}\otimes\boldsymbol{I})^{-1}\ \boldsymbol{x}(t)\\
&= (\boldsymbol{N}\boldsymbol{T}\otimes\boldsymbol{I})\ \tilde{\boldsymbol{x}}(t)\\
&= \left[\left(\ 0 \mid \hat{\boldsymbol{N}}\ \right)\otimes\boldsymbol{I}\right]\operatorname{diag}\left(\mathrm{e}^{(\boldsymbol{A}-\lambda_i\,\boldsymbol{B}\boldsymbol{K})\,t}\right)\tilde{\boldsymbol{x}}(0)\\
&= \left(\hat{\boldsymbol{N}}\otimes\boldsymbol{I}\right)\begin{pmatrix}\mathrm{e}^{(\boldsymbol{A}-\lambda_2\,\boldsymbol{B}\boldsymbol{K})\,t}\tilde{\boldsymbol{x}}_2(0)\\ \mathrm{e}^{(\boldsymbol{A}-\lambda_3\,\boldsymbol{B}\boldsymbol{K})\,t}\tilde{\boldsymbol{x}}_3(0)\\ \vdots\\ \mathrm{e}^{(\boldsymbol{A}-\lambda_N\,\boldsymbol{B}\boldsymbol{K})\,t}\tilde{\boldsymbol{x}}_N(0)\end{pmatrix}.
\end{aligned} \tag{A.6}$$

Since $\hat{\boldsymbol{N}}$ is an invertible matrix, the last investigation shows that the networked multi-agent system (3.9) is synchronized if and only if the matrices $\tilde{\boldsymbol{A}}_i = \boldsymbol{A} - \lambda_i\,\boldsymbol{B}\boldsymbol{K}$, $(i = 2, 3, \ldots, N)$ are stable. □

A.2 Proof of Lemma 3.3

Proof. The gradient $\partial\tilde{J}(\boldsymbol{K})/\partial\boldsymbol{K}$ of the objective function in eqn. (3.50) is obtained by application of the calculus of variations. The proof is divided into the following three steps.

1. Determination of the first variation

The variation of the objective function $\tilde{J}(\epsilon) = \tilde{J}(\boldsymbol{K} + \epsilon\,\boldsymbol{\eta})$ in eqn. (3.50) shows that

$$\tilde{J}(\epsilon) = \operatorname{tr}\left(\int_0^\infty \boldsymbol{\Phi}^T(t)\left((\boldsymbol{L}\otimes\boldsymbol{Q}) + (\boldsymbol{L}^2\otimes\boldsymbol{K}^T\boldsymbol{K}) + 2\epsilon\,(\boldsymbol{L}\otimes\boldsymbol{K}^T)\,\hat{\boldsymbol{\eta}} + \epsilon^2\,\hat{\boldsymbol{\eta}}^T\hat{\boldsymbol{\eta}}\right)\boldsymbol{\Phi}(t)\,\mathrm{d}t\right),$$

with

$$\boldsymbol{\Phi}(t) = \mathrm{e}^{\left((\boldsymbol{I}\otimes\boldsymbol{A}) - (\boldsymbol{L}\otimes\boldsymbol{B}\boldsymbol{K}) - \epsilon\,(\boldsymbol{L}\otimes\boldsymbol{B}\,\boldsymbol{\eta})\right)t}, \tag{A.7}$$

$\epsilon \in \mathbb{R}$, $\boldsymbol{\eta} \in \mathbb{R}^{m \times n}$ and $\hat{\boldsymbol{\eta}} = \boldsymbol{L} \otimes \boldsymbol{\eta}$. Since "high-order terms" approach 0 faster than ϵ approaches 0, it is obvious that only first order terms in ϵ are of interest. Hence,

$$\tilde{J}(\epsilon) = \operatorname{tr}\left(\int_0^\infty \boldsymbol{\Phi}^T(t) \Big(\underbrace{(\boldsymbol{L} \otimes \boldsymbol{Q})}_{\hat{\boldsymbol{Q}}} + \underbrace{(\boldsymbol{L}^2 \otimes \boldsymbol{K}^T \boldsymbol{K})}_{\hat{\boldsymbol{K}}^T \hat{\boldsymbol{K}}} + 2\epsilon \left(\boldsymbol{L} \otimes \boldsymbol{K}^T\right) \hat{\boldsymbol{\eta}} \Big) \boldsymbol{\Phi}(t) \, \mathrm{d}t \right) \tag{A.8}$$

holds.

Since the matrix exponential function (A.7) can be interpreted as the solution of the differential equation

$$\begin{aligned} \dot{\boldsymbol{\Phi}}(t) &= \Big((\boldsymbol{I} \otimes \boldsymbol{A}) - (\boldsymbol{L} \otimes \boldsymbol{B}\boldsymbol{K}) - \epsilon (\boldsymbol{L} \otimes \boldsymbol{B}\boldsymbol{\eta}) \Big) \boldsymbol{\Phi}(t), \qquad \boldsymbol{\Phi}(0) = \boldsymbol{I}, \\ &= \underbrace{\Big((\boldsymbol{I} \otimes \boldsymbol{A}) - (\boldsymbol{L} \otimes \boldsymbol{B}\boldsymbol{K}) \Big)}_{\tilde{\boldsymbol{A}}} \boldsymbol{\Phi}(t) - \epsilon (\boldsymbol{L} \otimes \boldsymbol{B}\boldsymbol{\eta}) \boldsymbol{\Phi}(t) \end{aligned}$$

$\boldsymbol{\Phi}(t)$ can be expressed to first order in ϵ by

$$\begin{aligned} \boldsymbol{\Phi}(t) &= \mathrm{e}^{\tilde{\boldsymbol{A}} t} - \epsilon \underbrace{\int_0^t \mathrm{e}^{\tilde{\boldsymbol{A}} (t-\tau)} (\boldsymbol{L} \otimes \boldsymbol{B}\boldsymbol{\eta}) \boldsymbol{\Phi}(\tau) \, \mathrm{d}\tau}_{\boldsymbol{\Psi}(t)} \\ &= \mathrm{e}^{\tilde{\boldsymbol{A}} t} - \epsilon \int_0^t \mathrm{e}^{\tilde{\boldsymbol{A}} (t-\tau)} (\boldsymbol{L} \otimes \boldsymbol{B}\boldsymbol{\eta}) \mathrm{e}^{\tilde{\boldsymbol{A}} \tau} \mathrm{d}\tau + \xcancel{\epsilon^2 \int_0^t \mathrm{e}^{\tilde{\boldsymbol{A}} (t-\tau)} (\boldsymbol{L} \otimes \boldsymbol{B}\boldsymbol{\eta}) \boldsymbol{\Psi}(\tau) \, \mathrm{d}\tau} \\ &= \mathrm{e}^{\tilde{\boldsymbol{A}} t} - \epsilon \tilde{\boldsymbol{\Psi}}(t) \end{aligned} \tag{A.9}$$

with

$$\tilde{\boldsymbol{\Psi}}(t) = \int_0^t \mathrm{e}^{\tilde{\boldsymbol{A}} (t-\tau)} (\boldsymbol{L} \otimes \boldsymbol{B}\boldsymbol{\eta}) \mathrm{e}^{\tilde{\boldsymbol{A}} \tau} \mathrm{d}\tau.$$

Substituting (A.9) in (A.8) yields

$$
\begin{aligned}
\tilde{J}(\epsilon) &= \mathrm{tr}\left(\int_0^\infty \left(\mathrm{e}^{\tilde{\boldsymbol{A}}t} - \epsilon\,\tilde{\boldsymbol{\Psi}}(t)\right)^T \left(\hat{\boldsymbol{Q}} + \hat{\boldsymbol{K}}^T\hat{\boldsymbol{K}} + 2\,\epsilon\,\hat{\boldsymbol{K}}^T\hat{\boldsymbol{\eta}}\right)\left(\mathrm{e}^{\tilde{\boldsymbol{A}}t} - \epsilon\,\tilde{\boldsymbol{\Psi}}(t)\right)\mathrm{d}t\right) \\
&= \underbrace{\mathrm{tr}\left(\int_0^\infty \mathrm{e}^{\tilde{\boldsymbol{A}}^T t}\left(\hat{\boldsymbol{Q}} + \hat{\boldsymbol{K}}^T\hat{\boldsymbol{K}}\right)\mathrm{e}^{\tilde{\boldsymbol{A}}t}\mathrm{d}t\right)}_{\tilde{J}(\boldsymbol{K})} + 2\,\epsilon\,\mathrm{tr}\left(\int_0^\infty \mathrm{e}^{\tilde{\boldsymbol{A}}^T t}\hat{\boldsymbol{K}}^T\hat{\boldsymbol{\eta}}\,\mathrm{e}^{\tilde{\boldsymbol{A}}t}\mathrm{d}t\right) \\
&\quad - \epsilon\,\mathrm{tr}\left(\int_0^\infty \tilde{\boldsymbol{\Psi}}^T(t)\left(\hat{\boldsymbol{Q}} + \hat{\boldsymbol{K}}^T\hat{\boldsymbol{K}} + 2\,\epsilon\,\hat{\boldsymbol{K}}^T\hat{\boldsymbol{\eta}}\right)\left(\mathrm{e}^{\tilde{\boldsymbol{A}}t} - \epsilon\,\tilde{\boldsymbol{\Psi}}(t)\right)\mathrm{d}t\right) \\
&\quad - \epsilon\,\mathrm{tr}\left(\int_0^\infty \mathrm{e}^{\tilde{\boldsymbol{A}}^T t}\left(\hat{\boldsymbol{Q}} + \hat{\boldsymbol{K}}^T\hat{\boldsymbol{K}} + 2\,\epsilon\,\hat{\boldsymbol{K}}^T\hat{\boldsymbol{\eta}}\right)\tilde{\boldsymbol{\Psi}}(t)\mathrm{d}t\right).
\end{aligned}
$$

The first variation can now be obtained by the derivative of $\tilde{J}(\epsilon) = J(\boldsymbol{K} + \epsilon\,\boldsymbol{\eta})$ with respect to ϵ evaluated at $\epsilon = 0$:

$$
\begin{aligned}
\delta\tilde{J}\Big|_{\boldsymbol{K}}(\boldsymbol{\eta}) &= \lim_{\epsilon \to 0} \frac{\tilde{J}(\boldsymbol{K} + \epsilon\,\boldsymbol{\eta}) - \tilde{J}(\boldsymbol{K})}{\epsilon} \\
&= 2\,\mathrm{tr}\left(\int_0^\infty \mathrm{e}^{\tilde{\boldsymbol{A}}^T t}\hat{\boldsymbol{K}}^T\hat{\boldsymbol{\eta}}\,\mathrm{e}^{\tilde{\boldsymbol{A}}t} - \mathrm{e}^{\tilde{\boldsymbol{A}}^T t}\left(\hat{\boldsymbol{Q}} + \hat{\boldsymbol{K}}^T\hat{\boldsymbol{K}}\right)\tilde{\boldsymbol{\Psi}}(t)\,\mathrm{d}t\right).
\end{aligned}
\tag{A.10}
$$

2. Decomposition of the first variation

Since the trace of a matrix is invariant under a similarity transformation with the orthogonal matrix $\tilde{\boldsymbol{T}} = (\boldsymbol{T} \otimes \boldsymbol{I})$, the right hand side of the eqn. (A.10) can be decomposed into a block diagonal form. The transformation of the exponential function yields

$$
\begin{aligned}
(\boldsymbol{T} \otimes \boldsymbol{I})^{-1}\,\mathrm{e}^{\tilde{\boldsymbol{A}}t}\,(\boldsymbol{T} \otimes \boldsymbol{I}) &= \mathrm{e}^{\left((\boldsymbol{I} \otimes \boldsymbol{A}) - (\boldsymbol{T}^{-1}\boldsymbol{L}\boldsymbol{T} \otimes \boldsymbol{B}\boldsymbol{K})\right)t} \\
&= \mathrm{e}^{\left((\boldsymbol{I} \otimes \boldsymbol{A}) - (\mathrm{diag}(\lambda_i) \otimes \boldsymbol{B}\boldsymbol{K})\right)t} \\
&= \mathrm{diag}\left(\mathrm{e}^{(\boldsymbol{A} - \lambda_i\,\boldsymbol{B}\,\boldsymbol{K})\,t}\right)
\end{aligned}
$$

and the transformations of the remaining terms of eqn. (A.10)

$$\begin{aligned}(\boldsymbol{T}\otimes\boldsymbol{I})^{-1}\underbrace{(\boldsymbol{L}\otimes\boldsymbol{K})^{T}(\boldsymbol{L}\otimes\boldsymbol{\eta})}_{\hat{\boldsymbol{K}}^{T}\hat{\boldsymbol{\eta}}}(\boldsymbol{T}\otimes\boldsymbol{I}) &= \left(\boldsymbol{T}^{-1}\boldsymbol{L}^{2}\boldsymbol{T}\otimes\boldsymbol{K}^{T}\boldsymbol{\eta}\right)\\ &= \left(\mathrm{diag}\left(\lambda_i^2\right)\otimes\boldsymbol{K}^{T}\boldsymbol{\eta}\right)\\ &= \mathrm{diag}\left(\lambda_i^2\,\boldsymbol{K}^{T}\boldsymbol{\eta}\right),\end{aligned}$$

$$(\boldsymbol{T}\otimes\boldsymbol{I})^{-1}\underbrace{\Big((\boldsymbol{L}\otimes\boldsymbol{Q})+\left(\boldsymbol{L}^{2}\otimes\boldsymbol{K}^{T}\boldsymbol{K}\right)\Big)}_{\hat{\boldsymbol{Q}}+\hat{\boldsymbol{K}}^{T}\hat{\boldsymbol{K}}}(\boldsymbol{T}\otimes\boldsymbol{I}) = \mathrm{diag}\left(\lambda_i\,\boldsymbol{Q}+\lambda_i^2\,\boldsymbol{K}^{T}\boldsymbol{K}\right)$$

and

$$(\boldsymbol{T}\otimes\boldsymbol{I})^{-1}\,\tilde{\boldsymbol{\Psi}}(t)\,(\boldsymbol{T}\otimes\boldsymbol{I}) = \mathrm{diag}\underbrace{\left(\int_0^t \mathrm{e}^{(\boldsymbol{A}-\lambda_i\,\boldsymbol{B}\,\boldsymbol{K})\,(t-\tau)}\lambda_i\,\boldsymbol{B}\,\boldsymbol{\eta}\,\mathrm{e}^{(\boldsymbol{A}-\lambda_i\,\boldsymbol{B}\,\boldsymbol{K})\,\tau}\mathrm{d}\tau\right)}_{\boldsymbol{\Psi_i}(t)}.$$

respectively.

Finally, the transformation of eqn. (A.10) is summarized as

$$\delta\tilde{J}\Big|_{\boldsymbol{K}}(\boldsymbol{\eta}) = 2\,\mathrm{tr}\left(\mathrm{diag}\left(\int_0^\infty \mathrm{e}^{\tilde{\boldsymbol{A}}_i^T t}\lambda_i^2\,\boldsymbol{K}^{T}\boldsymbol{\eta}\,\mathrm{e}^{\tilde{\boldsymbol{A}}_i\,t} - \mathrm{e}^{\tilde{\boldsymbol{A}}_i^T t}\left(\lambda_i\,\boldsymbol{Q}+\lambda_i^2\boldsymbol{K}^{T}\boldsymbol{K}\right)\boldsymbol{\Psi_i}(t),\mathrm{d}t\right)\right), \tag{A.11}$$

where $\tilde{\boldsymbol{A}}_i = \boldsymbol{A}-\lambda_i\,\boldsymbol{B}\,\boldsymbol{K}$, $(i = 2,\,3,\,\ldots,\,N)$. Since the trace of a block-diagonal matrix is the sum of the traces of its blocks, eqn. (A.11) can be reduced to

$$\delta\tilde{J}\Big|_{\boldsymbol{K}}(\boldsymbol{\eta}) = 2\sum_{i=2}^{N}\mathrm{tr}\left(\int_0^\infty \mathrm{e}^{\tilde{\boldsymbol{A}}_i^T t}\lambda_i^2\,\boldsymbol{K}^{T}\boldsymbol{\eta}\,\mathrm{e}^{\tilde{\boldsymbol{A}}_i\,t} - \mathrm{e}^{\tilde{\boldsymbol{A}}_i^T t}\left(\lambda_i\,\boldsymbol{Q}+\lambda_i^2\,\boldsymbol{K}^{T}\boldsymbol{K}\right)\boldsymbol{\Psi_i}(t),\mathrm{d}t\right). \tag{A.12}$$

3. Derivation of the necessary condition

By application of elementary properties of the trace operator it is possible to shift $\boldsymbol{\eta}$ to the left side of eqn. (A.12), which results in the following equations

$$\delta\tilde{J}\Big|_{\boldsymbol{K}}(\boldsymbol{\eta}) = \operatorname{tr}\left(\boldsymbol{\eta}^T 2 \sum_{i=2}^{N}\left[\int_0^\infty \lambda_i^2 \boldsymbol{K}\mathrm{e}^{\tilde{\boldsymbol{A}}_i t}\mathrm{e}^{\tilde{\boldsymbol{A}}_i^T t}\mathrm{d}t - \Gamma_i\right]\right) \tag{A.13}$$

with

$$\Gamma_i = \int_0^\infty \int_0^t \lambda_i \boldsymbol{B}^T \mathrm{e}^{\tilde{\boldsymbol{A}}_i^T (t-\tau)} \left(\lambda_i \boldsymbol{Q} + \lambda_i^2 \boldsymbol{K}^T\boldsymbol{K}\right) \mathrm{e}^{\tilde{\boldsymbol{A}}_i t}\mathrm{e}^{\tilde{\boldsymbol{A}}_i^T \tau}\mathrm{d}\tau\,\mathrm{d}t.$$

Interchanging the order of integration, shows that

$$\Gamma_i = \int_0^\infty \int_\tau^\infty \lambda_i \boldsymbol{B}^T \mathrm{e}^{\tilde{\boldsymbol{A}}_i^T (t-\tau)} \left(\lambda_i \boldsymbol{Q} + \lambda_i^2 \boldsymbol{K}^T\boldsymbol{K}\right) \mathrm{e}^{\tilde{\boldsymbol{A}}_i t}\mathrm{e}^{\tilde{\boldsymbol{A}}_i^T \tau}\mathrm{d}t\,\mathrm{d}\tau \tag{A.14}$$

holds. With the substitution $\kappa = t - \tau$, $\mathrm{d}t = \mathrm{d}\kappa$, it follows from (A.14) that

$$\begin{aligned}
\Gamma_i &= \int_0^\infty \int_0^\infty \lambda_i \boldsymbol{B}^T \mathrm{e}^{\tilde{\boldsymbol{A}}_i^T \kappa} \left(\lambda_i \boldsymbol{Q} + \lambda_i^2 \boldsymbol{K}^T\boldsymbol{K}\right) \mathrm{e}^{\tilde{\boldsymbol{A}}_i \kappa}\,\mathrm{e}^{\tilde{\boldsymbol{A}}_i \tau}\,\mathrm{e}^{\tilde{\boldsymbol{A}}_i^T \tau}\,\mathrm{d}\kappa\,\mathrm{d}\tau \\
&= \lambda_i \boldsymbol{B}^T \int_0^\infty \mathrm{e}^{\tilde{\boldsymbol{A}}_i^T \kappa} \left(\lambda_i \boldsymbol{Q} + \lambda_i^2 \boldsymbol{K}^T\boldsymbol{K}\right) \mathrm{e}^{\tilde{\boldsymbol{A}}_i \kappa}\,\mathrm{d}\kappa \int_0^\infty \mathrm{e}^{\tilde{\boldsymbol{A}}_i \tau}\,\mathrm{e}^{\tilde{\boldsymbol{A}}_i^T \tau}\,\mathrm{d}\tau \\
&= \lambda_i \boldsymbol{B}^T \boldsymbol{P}_i \boldsymbol{X}_i
\end{aligned} \tag{A.15}$$

can be obtained by a product of the matrices

$$\boldsymbol{P}_i = \int_0^\infty \mathrm{e}^{\tilde{\boldsymbol{A}}_i^T \kappa} \left(\lambda_i \boldsymbol{Q} + \lambda_i^2 \boldsymbol{K}^T\boldsymbol{K}\right) \mathrm{e}^{\tilde{\boldsymbol{A}}_i \kappa}\,\mathrm{d}\kappa \tag{A.16}$$

and

$$\boldsymbol{X}_i = \int_0^\infty \mathrm{e}^{\tilde{\boldsymbol{A}}_i \tau}\,\mathrm{e}^{\tilde{\boldsymbol{A}}_i^T \tau}\mathrm{d}\tau. \tag{A.17}$$

The first-order necessary condition for optimality is

$$\delta\tilde{J}\Big|_{\boldsymbol{K}}(\boldsymbol{\eta}) = 0, \quad \forall\, \boldsymbol{\eta} \in \mathbb{R}^{m\times n}.$$

This condition is satisfied with eqns. (A.13), (A.15) and (A.16) – (A.17) if and only if

$$0 = 2\sum_{i=2}^{N} \left(\lambda_i^2 \, \boldsymbol{K} \, \boldsymbol{X}_i - \lambda_i \, \boldsymbol{B}^T \boldsymbol{P}_i \, \boldsymbol{X}_i\right) \tag{A.18}$$

holds. The right-hand side of eqn. (A.18) represents the gradient $\nabla J(\boldsymbol{K})$. Note that $\lambda_1 = 0$ and that $\boldsymbol{X}_i$ and $\boldsymbol{P}_i$ are symmetric positive definite solutions of the following Lyapunov equations

$$(\boldsymbol{A} - \lambda_i \, \boldsymbol{B} \, \boldsymbol{K})^T \, \boldsymbol{P}_i + \boldsymbol{P}_i \, (\boldsymbol{A} - \lambda_i \, \boldsymbol{B} \, \boldsymbol{K}) + \lambda_i \, \boldsymbol{Q} + \lambda_i^2 \, \boldsymbol{K}^T \boldsymbol{K} = \boldsymbol{O}$$
$$(\boldsymbol{A} - \lambda_i \, \boldsymbol{B} \, \boldsymbol{K}) \, \boldsymbol{X}_i + \boldsymbol{X}_i \, (\boldsymbol{A} - \lambda_i \, \boldsymbol{B} \, \boldsymbol{K})^T + \boldsymbol{I} = \boldsymbol{O}.$$

□

A.3 Proof of Lemma 3.4

Proof. The linearization of $\tilde{J}(\boldsymbol{K} - a\nabla \tilde{J}(\boldsymbol{K}))$ around $a = 0$ shows that

$$\tilde{J}(\boldsymbol{K} - a\nabla \tilde{J}(\boldsymbol{K})) = \tilde{J}(\boldsymbol{K}) - a \operatorname{tr}\left(\nabla^T \tilde{J}(\boldsymbol{K}) \, \nabla \tilde{J}(\boldsymbol{K})\right)$$

holds. Since $\boldsymbol{H} = \left(\nabla \tilde{J}(\boldsymbol{K})\right)^T \nabla \tilde{J}(\boldsymbol{K}) \succeq 0$, it follows that

$$\operatorname{tr}(\boldsymbol{H}) = \sum_{i=1}^{n} \lambda_i \, (\boldsymbol{H}) \geq 0.$$

This implies $a > 0$ to exist such that $\tilde{J}(\boldsymbol{K} - a\nabla \tilde{J}(\boldsymbol{K})) \leq \tilde{J}(\boldsymbol{K})$. □

B Proofs of Chapter 4

B.1 Proof of Theorem 4.1

Proof. The overall synchronization error $\bar{e}(t) \circ\!\!-\!\!\bullet \bar{\boldsymbol{E}}(s)$ where

$$\bar{\boldsymbol{E}}(s) = \begin{pmatrix} Y_1(s) - Y_2(s) \\ Y_1(s) - Y_3(s) \\ \vdots \\ Y_1(s) - Y_N(s) \end{pmatrix} = \underbrace{\begin{pmatrix} 1 & -1 & 0 & \cdots & 0 \\ 1 & 0 & -1 & \ddots & \vdots \\ \vdots & \vdots & \ddots & \ddots & 0 \\ 1 & 0 & \cdots & 0 & -1 \end{pmatrix}}_{\boldsymbol{N}} \boldsymbol{Y}(s)$$

can by virtue of Lemma 4.1 be formulated as

$$\begin{aligned} \bar{\boldsymbol{E}}(s) &= \boldsymbol{N}\, (\boldsymbol{I} + G(s)\, K(s)\, \boldsymbol{L})^{-1}\, \boldsymbol{D}(s) \\ &= \boldsymbol{N}\boldsymbol{T}\, \left(\boldsymbol{I} + G(s)\, K(s)\, \boldsymbol{T}^{-1}\, \boldsymbol{L}\boldsymbol{T}\right)^{-1}\, \boldsymbol{T}^{-1}\, \boldsymbol{D}(s) \\ &= \boldsymbol{N}\boldsymbol{T}\, \left(\boldsymbol{I} + G(s)\, K(s)\, \operatorname*{diag}_{i=1}^{N}(\lambda_i)\right)^{-1}\, \boldsymbol{T}^{-1}\, \boldsymbol{D}(s) \\ &= \boldsymbol{N}\boldsymbol{T}\, \operatorname*{diag}_{i=1}^{N}\left(\frac{1}{1 + \lambda_i\, G(s)\, K(s)}\right)\, \boldsymbol{T}^{-1}\, \boldsymbol{D}(s) \end{aligned} \qquad \text{(B.1)}$$

where $\boldsymbol{T}^{-1}\, \boldsymbol{L}\boldsymbol{T} = \operatorname{diag}(\lambda_i)$ with $\lambda_1 = 0 \leq \lambda_2 \leq \ldots \leq \lambda_N$ holds. The fact that the row sum of the matrix $\boldsymbol{N}$ is zero and the first eigenvector of the Laplacian matrix $\boldsymbol{L}$ is given by $\boldsymbol{t}_1 = \mathbb{1}/\sqrt{N}$ allows to reduce eqn. (B.1) to its bijective basis. Since

$$\boldsymbol{N}\boldsymbol{T} = \boldsymbol{N}\underbrace{\left(\tfrac{1}{\sqrt{N}}\mathbb{1} \quad \boldsymbol{t}_2 \quad \cdots \quad \boldsymbol{t}_N\right)}_{\boldsymbol{T}} = \left(\, \boldsymbol{0} \,\middle|\, \hat{\boldsymbol{N}} \,\right)$$

it follows directly from eqn. (B.1) that

$$\bar{\boldsymbol{E}}(s) = \hat{\boldsymbol{N}} \operatorname*{diag}_{i=2}^{N} \left(\frac{1}{1 + \lambda_i \, G(s) \, K(s)} \right) \begin{pmatrix} \tilde{d}_2 \\ \vdots \\ \tilde{d}_N \end{pmatrix} \frac{1}{s}$$

with $\begin{pmatrix} \tilde{d}_1 & \tilde{d}_2 & \cdots & \tilde{d}_N \end{pmatrix}^T = \boldsymbol{T}^{-1} \boldsymbol{d}$ holds.

Since $\operatorname{rank}(\boldsymbol{N}\,\boldsymbol{T}) = \operatorname{rank}(\boldsymbol{N}) = N - 1$ it is easy to see that the matrix $\hat{\boldsymbol{V}} \in \mathbb{R}^{N-1 \times N-1}$ is invertible and hence the overall system (4.11) with respect to eqn. (4.1) asymptotically synchronized if and only if all poles of the uncoupled dynamics

$$\tilde{S}_i(s) = \frac{1}{1 + \lambda_i \, G(s) \, K(s)}, \quad i = 2, 3, \ldots, N$$

have a negative real part and $\lim_{t \to \infty} \bar{\boldsymbol{e}}(t) = \lim_{s \to 0} s \, \bar{\boldsymbol{E}}(s) = 0$ holds. Finally, the requirement $\lim_{s \to 0} s \, \bar{\boldsymbol{E}}(s) = 0$ is fulfilled if and only if the open-loop transfer function $G_0(s) = G(s) \, K(s)$ includes integrator dynamics. □

B.2 Proof of Theorem 4.3

Proof. According to the statements of the lemma, the proof is achieved separately for underdamped and overdamped transient responses and for the transition between them.

Underdamped transient responses

The underdamped transient response as defined in (4.26) for $d_i < 1$, is bounded by its exponential envelope

$$|h_i(t)| = \frac{1}{\sqrt{1 - d_i^2}} \mathrm{e}^{-\omega_0 \, d \, t} \left| \sin\left(\omega_0 \sqrt{\lambda_i - d^2} \, t + \operatorname{acos}(d_i)\right) \right| \le \hat{h}_i(t).$$

The envelope

$$\hat{h}_i(t) = \frac{\mathrm{e}^{-\omega_0 \, d \, t}}{\sqrt{1 - d_i^2}}$$

monotonically increases as d_i approaches the value one. Consequently, the envelopes which correspond to larger Laplacian eigenvalues are bounded by those which correspond to smaller ones, which proves the statement (4.31).

Overdamped transient responses

The overdamped transient response has been defined in (4.26) for $d_i > 1$, where the poles are given by (4.25). The stated relation (4.29) holds in general if the derivative

$$h_i'(t) = \frac{\mathrm{d}\, h_i(t)}{\mathrm{d}\, \lambda_i}$$

is non-positive for all parametrizations. The derivative of $h_i(t)$ for $d_i > 1$ with respect to λ_i is given by:

$$\begin{aligned}\frac{\mathrm{d}\, h_i(t)}{\mathrm{d}\, \lambda_i} =& \frac{-2\, s_{1i}'}{(s_{1i} - s_{2i})^2} \left(s_{1i}\, \mathrm{e}^{s_{2i}\, t} - s_{2i}\, \mathrm{e}^{s_{1i}\, t} \right) + \ldots \\ & \frac{s_{1i}'}{s_{1i} - s_{2i}} \left(\mathrm{e}^{s_{1i}\, t} + \mathrm{e}^{s_{2i}\, t} - s_{2i}\, t\, \mathrm{e}^{s_{1i}\, t} - s_{1i}\, t\, \mathrm{e}^{s_{2i}\, t} \right),\end{aligned}$$

where

$$s_{1i}' = \frac{\mathrm{d}\, s_{1i}}{\mathrm{d}\, \lambda_i} < 0.$$

Multiplying the statement to be proved $h_i'(t) \overset{!}{\leq} 0$ with

$$\mathrm{e}^{-s_{1i}\, t}\, \frac{(s_{1i} - s_{2i})^2}{s_{1i}'} < 0$$

yields

$$-2 \left(s_{1i}\, \mathrm{e}^{\Delta s_i\, t} - s_{2i} \right) + (s_{1i} - s_{2i}) \left(\mathrm{e}^{\Delta s_i\, t} + 1 - s_{2i}\, t - s_{1i}\, t\, \mathrm{e}^{\Delta s_i\, t} \right) \overset{!}{\geq} 0, \qquad \text{(B.2)}$$

where $\Delta s_i = s_{2i} - s_{1i}$. The following equation is obtained by separating the exponential function in (B.2):

$$\mathrm{e}^{\Delta s_i\, t} \overset{!}{\geq} \frac{-2\, s_{2i} - (s_{1i} - s_{2i})\, (1 - s_{2i}\, t)}{-2\, s_{1i} + (s_{1i} - s_{2i})\, (1 - s_{1i}\, t)}.$$

Since $\mathrm{e}^{\Delta s_i\, t} \geq 1$, the statement is proved if

$$1 \overset{!}{\geq} \frac{-2\, s_{2i} - (s_{1i} - s_{2i})\, (1 - s_{2i}\, t)}{-2\, s_{1i} + (s_{1i} - s_{2i})\, (1 - s_{1i}\, t)}$$

holds. Since the denominator is positive, it is easy to verify from

$$-\, (s_{1i} - s_{2i})\, (s_{1i} + s_{2i})\, t \overset{!}{\geq} 0$$

that $h_i'(t) \leq 0$ as $s_{1i} + s_{2i} \leq 0$. This inequality proves statement (4.29).

Transition between the over- and underdamped case

The underdamped transient responses are bounded by overdamped transient responses defined in (4.26) if

$$\frac{1}{\sqrt{1-d_{i+1}^2}}\,\mathrm{e}^{-\omega_0\,d\,t}\sin(\alpha) \overset{!}{\leq} \frac{s_{1i}\,\mathrm{e}^{s_{2i}\,t} - s_{2i}\,\mathrm{e}^{s_{1i}\,t}}{s_{1i} - s_{2i}}$$

with

$$\alpha = \omega_0\sqrt{\lambda_{i+1} - d^2}\,t + \operatorname{acos}(d_{i+1})$$

holds. A rearrangement of the equation and a substitution with (4.25) gives

$$\sin(\alpha) \overset{!}{\leq} \frac{\sqrt{1-d_{i+1}^2}}{s_{1i} - s_{2i}}\left(s_{1i}\mathrm{e}^{-\mu\,t} - s_{2i}\mathrm{e}^{\mu\,t}\right),$$

where $\mu = \omega_0\sqrt{d^2 - \lambda_i}$. Since $\sin(\alpha) \leq 1$, the lemma is proved if

$$1 \overset{!}{\leq} \frac{\sqrt{1-d_{i+1}^2}}{s_{1i} - s_{2i}}\left(s_{1i}\mathrm{e}^{-\mu\,t} - s_{2i}\mathrm{e}^{\mu\,t}\right) \tag{B.3}$$

holds. The first term of right-hand side of eqn. (B.3) is positive. Since, the minimum value of the sum in eqn. (B.3) occurs at $t = 0$ it is obvious that eqn. (B.3) and hence, statement (4.30) holds if

$$1 \leq \sqrt{1-d_{i+1}^2},$$

which is fulfilled by assumption $d_{i+1} < 1$.

Finally, the lemma is proved by the summary of the proofs of each of the statements. □

C Proofs of Chapter 5

C.1 Proof of Theorem 5.1

Proof. With the state transformation

$$\tilde{\boldsymbol{x}}(k) = \left(\underbrace{\left(\begin{array}{c|ccc} 1 & 0 & \cdots & 0 \\ \hline 1 & & & \\ \vdots & & -\boldsymbol{I} & \\ 1 & & & \end{array} \right)}_{\boldsymbol{V}} \otimes \boldsymbol{I} \right) \boldsymbol{x}(k) \tag{C.1}$$

the solution (5.25) is obtained as

$$\begin{aligned} \tilde{\boldsymbol{x}}(\kappa P n) &= (\boldsymbol{V} \otimes \boldsymbol{I}) \left(\boldsymbol{T} \otimes \boldsymbol{A}^{Pn} \right)^{\kappa} (\boldsymbol{V}^2 \otimes \boldsymbol{I}) \, \boldsymbol{x}_0 \\ &= \left(\boldsymbol{V} \boldsymbol{T} \boldsymbol{V} \otimes \boldsymbol{A}^{Pn} \right)^{\kappa} (\boldsymbol{V} \otimes \boldsymbol{I}) \, \boldsymbol{x}_0 \\ &= \left(\left(\begin{array}{c|ccc} 1 & * & \cdots & * \\ \hline 0 & & & \\ \vdots & & \tilde{\boldsymbol{T}} & \\ 0 & & & \end{array} \right) \otimes \boldsymbol{A}^{Pn} \right)^{\kappa} (\boldsymbol{V} \otimes \boldsymbol{I}) \, \boldsymbol{x}_0 . \end{aligned} \tag{C.2}$$

Note that $\boldsymbol{V}^2 = \boldsymbol{I}$ holds. According to requirement (5.1), the agents are synchronized if and only if every entry of the error vector

$$\boldsymbol{e}(k) = (\boldsymbol{N} \otimes \boldsymbol{I}) \, \boldsymbol{x}(k) = \begin{pmatrix} \boldsymbol{x}_1(k) - \boldsymbol{x}_2(k) \\ \boldsymbol{x}_1(k) - \boldsymbol{x}_3(k) \\ \vdots \\ \boldsymbol{x}_1(k) - \boldsymbol{x}_N(k) \end{pmatrix} \tag{C.3}$$

converges asymptotically to zero. $\boldsymbol{N} = (\boldsymbol{0} \mid \boldsymbol{I}) \, \boldsymbol{V}$ is an $(N-1 \times N)$-matrix which by construction has a row sum equal to zero. Applying the transformation (C.1) and eqn. (C.2) to the error vector (C.3) yields

$$\begin{aligned}
\boldsymbol{e}(\kappa P n) &= (\boldsymbol{N} \otimes \boldsymbol{I})\,(\boldsymbol{V}^2 \otimes \boldsymbol{I})\,\boldsymbol{x}(\kappa P n) \\
&= (\boldsymbol{N}\boldsymbol{V} \otimes \boldsymbol{I})\,\tilde{\boldsymbol{x}}(\kappa P n) \\
&= \left[\begin{pmatrix} 0 & \boldsymbol{I} \end{pmatrix} \otimes \boldsymbol{I}\right] \tilde{\boldsymbol{x}}(\kappa P n) \\
&= \left(\begin{pmatrix} 0 & \tilde{\boldsymbol{T}}^{\kappa} \end{pmatrix} \otimes \boldsymbol{A}^{Pn\kappa}\right)(\boldsymbol{V} \otimes \boldsymbol{I})\,\boldsymbol{x}_0 \\
&= \left(\tilde{\boldsymbol{T}}^{\kappa} \begin{pmatrix} 0 & \boldsymbol{I} \end{pmatrix} \otimes \boldsymbol{A}^{Pn\kappa}\right)(\boldsymbol{V} \otimes \boldsymbol{I})\,\boldsymbol{x}_0 \\
&= \left(\tilde{\boldsymbol{T}} \otimes \boldsymbol{A}^{Pn}\right)^{\kappa} \left(\begin{pmatrix} 0 & \boldsymbol{I} \end{pmatrix} \boldsymbol{V} \otimes \boldsymbol{I}\right) \boldsymbol{x}_0 \\
&= \left(\tilde{\boldsymbol{T}} \otimes \boldsymbol{A}^{Pn}\right)^{\kappa} (\boldsymbol{N} \otimes \boldsymbol{I})\,\boldsymbol{x}_0 \\
&= \left(\tilde{\boldsymbol{T}} \otimes \boldsymbol{A}^{Pn}\right)^{\kappa} \boldsymbol{e}(0).
\end{aligned} \tag{C.4}$$

The investigations show that the overall closed-loop system (5.14) is synchronized if and only if all the eigenvalues of the matrix $\tilde{\boldsymbol{T}} \otimes \boldsymbol{A}^{Pn}$ are inside the unit circle of the complex plane, which is the case if and only if $\rho(\tilde{\boldsymbol{T}})\rho(\boldsymbol{A}^{Pn}) < 1$. The similarity transformation in (C.2) shows that the eigenvalues of $\tilde{\boldsymbol{T}}$ coincide with the eigenvalues of the matrix $\boldsymbol{T}$ with the exception of $\lambda_1 = 1$. Hence, $\rho(\tilde{\boldsymbol{T}}) = |\lambda_2|$ where $|\lambda_2|$ is the second largest magnitude among the eigenvalues of $\boldsymbol{T}$. The combination of these facts proves the theorem. □

C.2 Proof of Theorem 5.3

Proof. The proof of Theorem 5.3 is close to the proof of Theorem 5.1. The transformation of (5.42) with (C.1) gives

$$\begin{aligned}
(\boldsymbol{V} \otimes \boldsymbol{I})\,\mathrm{E}(\boldsymbol{x}(rn)) &= (\boldsymbol{V} \otimes \boldsymbol{I})\left(\bar{\boldsymbol{T}} \otimes \boldsymbol{A}^{n}\right)^{r} (\boldsymbol{V}^2 \otimes \boldsymbol{I})\,\boldsymbol{x}_0 \\
&= \left(\left(\begin{array}{c|ccc} 1 & * & \cdots & * \\ \hline 0 & & & \\ \vdots & & \tilde{\bar{\boldsymbol{T}}} & \\ 0 & & & \end{array}\right) \otimes \boldsymbol{A}^{n}\right)^{r} \tilde{\boldsymbol{x}}(0).
\end{aligned} \tag{C.5}$$

The zeros in (C.5) result from the property $\bar{\boldsymbol{T}}\mathbb{1} = \mathbb{1}$ of the stochastic matrix $\bar{\boldsymbol{T}}$. Considering the synchronization error $\tilde{\text{e}}(rn) = (\boldsymbol{N} \otimes \boldsymbol{I})\text{E}(\boldsymbol{x}(rn))$ together with eqn. (C.5) shows that

$$\begin{aligned}\tilde{\text{e}}(rn) &= (\boldsymbol{N} \otimes \boldsymbol{I})\ (\boldsymbol{V}^2 \otimes \boldsymbol{I})\,\text{E}(\boldsymbol{x}(rn))\\ &= (\boldsymbol{N}\boldsymbol{V} \otimes \boldsymbol{I})\ (\boldsymbol{V} \otimes \boldsymbol{I})\text{E}(\boldsymbol{x}(rn))\\ &= \left(\tilde{\bar{\boldsymbol{T}}} \otimes \boldsymbol{A}^n\right)^r \tilde{\text{e}}(0),\end{aligned} \tag{C.6}$$

holds. For the analysis of the convergence of the synchronization error $\tilde{\text{e}}(rn)$, eqn. (C.6) can be treated similarly to eqn. (C.4): $\tilde{\text{e}}(rn)$ converges to zero if and only if $\rho(\tilde{\bar{\boldsymbol{T}}})\rho(\boldsymbol{A}^n) < 1$. Note that $\rho(\tilde{\bar{\boldsymbol{T}}}) = \bar{\lambda}_2$ where $\bar{\lambda}_2$ is the second largest eigenvalue of the symmetric matrix $\tilde{\bar{\boldsymbol{T}}}$. □

D List of Symbols

General conventions

Description	Notation	Example
Scalars	small, italic	a, b
Vectors	small, bold, italic	$\boldsymbol{a}$, $\boldsymbol{b}$
Matrices	capital, bold, italic	$\boldsymbol{A}$, $\boldsymbol{B}$
Sets	capital, calligraphic	$\mathcal{N}$, $\mathcal{P}$

Operations and definitions

Symbol	Description
$(.)^{-1}$	Inverse of a matrix
$(.)^T$	Transpose of a vector or a matrix
$(.)_{ij}$	Entry of the i-th row and the j-th column of a matrix
$(.)_i$	The i-th element of a vector
$\Re\{.\}$	Real part of a complex number
$\Im\{.\}$	Imaginary part of a complex number
$\mathrm{tr}(.)$	Trace of a square matrix
$\mathrm{rank}\,(.)$	Rank of a square matrix
$\mathrm{diag}(.)$	Diagonal or block diagonal matrix
$\lambda_i(.)$	The i-th eigenvalue of a matrix with $\lvert\lambda_i(.)\rvert \leq \lvert\lambda_{i+1}(.)\rvert$
$\lambda_{\max}(.)$	Largest eigenvalue of a matrix
$\rho(.)$	Spectral radius of a matrix
$\lceil.\rceil$	Smallest integer larger than or equal to a scalar
$\mathrm{E}(.)$	Expected value of a discrete random variable
$\bullet\!\!-\!\!\circ$	Laplace transformation
$\lvert.\rvert$	Cardinality of a set or the absolute value of a scalar
$\lVert.\rVert$	Euclidean vector norm or spectral norm
$\lVert.\rVert_{\mathrm{F}}$	Frobenius norm of a matrix
$\otimes$	Kronecker product

∇	Gradient of a vector or matrix valued function
$\mathbb{R}$	Set of real numbers
$\mathbb{N}$	Set of natural numbers
$\mathbb{Z}$	Set of integers
$\mathbb{1}$	One vector
$\mathbf{0}$	Zero vector
$\succ$	Definiteness of a matrix
$\sigma(t)$	Step signal

Andrej Mosebach

Geburtsdatum:	16.06.1984
Geburtsort:	Ananjewo, Kirgistan

Bildungs- und Berufsweg

seit 11/2016	Ingenieur Vorentwicklung, Vorwerk Elektrowerke GmbH & Co. KG
07/2011 – 03/2016	Wissenschaftlicher Mitarbeiter, Lehrstuhl für Automatisierungstechnik und Prozessinformatik, Ruhr-Universität Bochum
10/2008 – 05/2011	Master of Science in Elektrotechnik und Informationstechnik, Studienschwerpunkt Automatisierungstechnik, Ruhr-Universität Bochum Masterarbeit: *Autonomie und Kooperation in Multiagentensystemen*
10/2005 – 10/2008	Bachelor of Engineering in Elektrotechnik für Energie, Licht und Automation, Fachhochschule Südwestfalen Hagen Bachelorarbeit: *Aufbau und Regelung eines inversen Pendels*
09/2002 – 06/2005	Gesamtschule Königsborn in Unna